# CAMBRIDGE STUDIES IN ECOLOGY

**The terrestrial invasion**

# The terrestrial invasion

## An ecophysiological approach to the origins of land animals

COLIN LITTLE

*Senior Lecturer, Department of Zoology*
*University of Bristol*

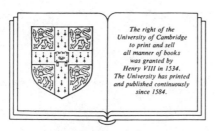

CAMBRIDGE UNIVERSITY PRESS

*Cambridge*

*New York   Port Chester*

*Melbourne   Sydney*

Published by the Press Syndicate of the University of Cambridge
The Pitt Building, Trumpington Street, Cambridge CB2 1RP
40 West 20th Street, New York, NY 10011, USA
10 Stamford Road, Oakleigh, Melbourne 3166, Australia

First published 1990

Printed in Great Britain at the University Press, Cambridge

*British Library cataloguing in publication data*

Little, Colin
The terrestrial invasion: an ecophysiological
approach to the origins of land animals.
1. Environment. Adaptation of animals
I. Title
591.5

*Library of Congress cataloguing in publication data*

Little, Colin, 1939–
The terrestrial invasion: an ecophysiological approach to the
origins of land animals / Colin Little.
     p.   cm.
ISBN 0-521-33447-0. – ISBN 0-521-33669-4 (pbk).
1. Evolution.   2. Adaptation (Biology)   3. Animal ecology.
I. Title
QH371.L58   1990
560'.45–dc20   89-17471 CIP

ISBN 0 521 33447 0  hardback
ISBN 0 521 33669 4  paperback

VN

To SERG, for all its support, in water as well as on land

# Contents

# Preface

This book is intended as an introduction to a very wide subject, the invasion of land by animal lines which originated in aquatic environments. In an earlier book, *The Colonisation of Land*, I discussed the phylogeny and physiology of terrestrial and semi-terrestrial animals in some detail. Here I have taken a more ecological approach, considering the various possible intermediate environments between water and land, and the ways in which animal lines have adapted to these. To do this I have adopted a 'case-history' approach, which has allowed me to select examples that seem to me to be representative. I have also dealt more extensively with the environment itself, and with the ways in which terrestrial communities have developed. This book is therefore in every way a *new* book, although some of the physiological information in the first book has been summarized in Chapters 4 and 9.

During the writing of the text, I have been fortunate to benefit from the advice of Dr R. S. K. Barnes, who originally suggested the overall plan, and who has made constructive comments on all the chapters. I thank him, and my wife P. E. Stirling, who has also read the entirety, and has helped to reduce it to English. I am grateful to Dr D. E. G. Briggs, whose advice has allowed me to re-draft and update Chapters 1 and 3. Lastly, I am lucky to be able to thank Mrs J. Ablett, who once again has struggled with my text to produce an excellent index.

Colin Little

Bristol

# Part I

## Introduction

In this introductory section, present-day terrestrial ecosystems are first placed against a background of gradual development over more than 400 million years (Chapter 1). To give some understanding of the problems faced by aquatic animal lines moving on to land during this period, Chapter 2 compares the properties of aquatic and terrestrial environments, and points out the extent of biological influence on terrestrial climates. Chapter 3 concludes the introduction by discussing the sources of evidence available to us in tracing the origins of terrestrial animals.

# 1

## Perspective

at successive periods of the past, the same area of land and
water has been inhabited by species of animals and plants even
more distinct than those which now people the antipodes, or
which now co-exist in the arctic, temperate, and tropical zones.
Sir Charles Lyell (1878) *The Student's Elements of Geology*, 3rd edn.
London: John Murray.

The oldest records of terrestrial animals date from the Silurian.
At this time there were relatives of the present-day myriapods on land. This
does not mean, of course, that there was no terrestrial fauna before the
Silurian, only that we have no record of it: the fossil record provides many
problems of interpretation, as will be explained in Chapter 3. Certainly,
however, the diversity of animal life on land was very low. After the
Silurian, in the Devonian, Carboniferous and subsequent periods, we have
records of the progressively increasing diversity of terrestrial animals (Fig.
1.1). Step by step, terrestrial ecosystems evolved until by Tertiary times
they became very similar to those of the present day. This invasion of the
land by primitively aquatic animals, and the adaptations and diversification of animals en route to land provide the theme for this book.

One of the major characteristics of terrestrial faunas is their narrow
phylogenetic base in comparison with that of aquatic faunas. While there
are nearly 30 free-living macrofaunal phyla extant today, 11 of these are
confined to the sea, and only 7 have truly terrestrial representatives (Table
1.1). Several more phyla, containing small forms, together with protists and
bacteria, are found in the soil. No phyla are restricted to fresh water, and
only one, the Onychophora, is restricted to land. This distribution has led
most authorities to consider that life originated in the sea, and that
freshwater and terrestrial forms evolved later (e.g. Pearse, 1936). An
alternative hypothesis was developed by Hinton & Blum (1965), who
suggested that life originated on land, but this has not received general

3

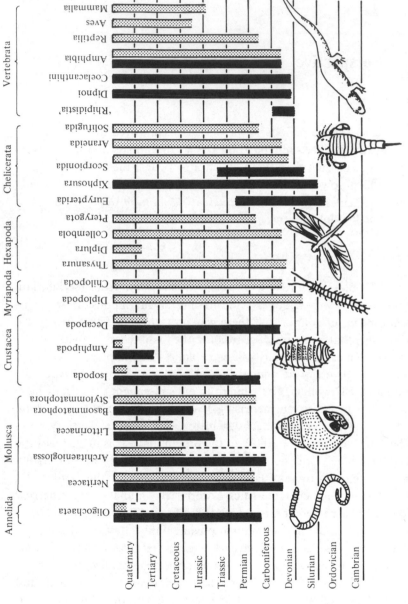

Fig. 1.1. The fossil record of some animal groups, greatly simplified. Aquatic records are shown as black bars, terrestrial records as stipple. Dotted lines show uncertain records. After Little (1983). Data taken from many authors, and modified by reference particularly to Rolfe (1985), Solem & Yochelson (1979) and D.E.G. Briggs (personal communication).

Table 1.1 *The distributions of animal phyla*

| | Sea | Fresh water | Soil | Land (above soil surface) |
|---|---|---|---|---|
| Protista | X | X | X | — |
| Porifera | X | X | — | — |
| Cnidaria | X | X | — | — |
| Platyhelminthes | X | X | X | X |
| Ctenophora | X | — | — | — |
| Nemertea | X | X | — | X |
| Rotifera | X | X | X | — |
| Gastrotricha | X | X | — | — |
| Kinorhyncha | X | — | — | — |
| Nematoda | X | X | X | — |
| Nematomorpha | X | X | — | — |
| Entoprocta | X | X | — | — |
| Annelida | X | X | X | X |
| Mollusca | X | X | X | X |
| Phoronida | X | — | — | — |
| Bryozoa | X | X | — | — |
| Brachiopoda | X | — | — | — |
| Sipunculida | X | — | — | — |
| Echiuroida | X | — | — | — |
| Priapulida | X | — | — | — |
| Tardigrada | X | X | X | — |
| Onychophora | — | — | X | X |
| Arthropoda | X | X | X | X |
| Echinodermata | X | — | — | — |
| Chaetognatha | X | — | — | — |
| Pogonophora | X | — | — | — |
| Hemichordata | X | — | — | — |
| Chordata | X | X | X | X |

acceptance. Wherever life evolved, most animal phyla probably developed in the sea, and only later moved on to land.

Before investigating the mechanisms involved in the transition from sea to land, and the routes taken by various animal lines, this introductory chapter attempts to give a picture of the time scale over which the invasion of the land occurred. Although much of the book then deals with the evolution of terrestrial *animals*, it is vital at this stage to consider the development of life on land as the development of a terrestrial *ecosystem*, in which green plants play an essential role. This is not the place for a comprehensive discussion of the origins of terrestrial plants, however, and good accounts of this have already been given (see Chaloner, 1970; Delevoryas, 1977; and Stewart, 1983). Fig. 1.2 shows the fossil record of

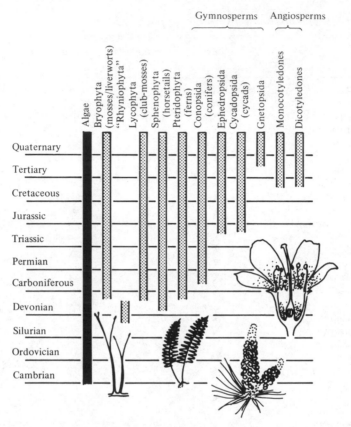

Fig. 1.2. The fossil record of some groups of plants. Aquatic records black, terrestrial records stipple. After Little (1983) from several authors.

major plant groups. The development of terrestrial ecosystems has been considered recently in more detail by Seldon & Edwards (1989) in relation to changes in the physico-chemical properties of the earth's surface. Their account gives more details of the fossil record of organisms than can be provided here, and should be consulted for further information. The physiological adaptations of plants to life on land have been detailed by Spicer (1989). The present account aims merely to provide an overall view as an introduction to later chapters.

## 1.1    The first life on land: Ordovician and Silurian ecosystems

Although the earliest indisputible terrestrial organisms date from the Silurian, circumstantial evidence suggests that there was some life on land earlier than this. Micro-organisms probably invaded the land during the Precambrian (Wright, 1985), and it is possible that multicellular plants evolved on land directly from unicellular terrestrial organisms, and not from multicellular aquatic plants (Stebbins & Hill, 1980). By the Ordovician, it is possible that algae and bryophytes had evolved on land, and formed a biologically active soil cover, although we have no fossil record of this invasion (Wright, 1985). It is also possible that some invertebrates had become terrestrial, although once again we have no direct fossil remains of them. In support of these suggestions, the study of fossil soils of the Ordovician (Retallack, 1985) has demonstrated the presence of burrows, possibly made by invertebrates, and other fossil structures which may represent traces of root systems or fungal mycelia. Fossil spore tetrads have also been recovered from late Ordovician deposits, so that there may have been a well-developed land flora of non-vascular plants. J. Gray (1985a,b) has argued that these plants were widespread, but like the bryophytes may have been lacking in morphological adaptations allowing them to control their water economy. Other authors (see the discussion following Gray, 1985a) have contended that the spores may not necessarily belong to land plants.

At present, any reconstruction of Ordovician land surfaces must be purely speculative, but there does seem a possibility that at least in the region of water bodies there was a fringing region of non-vascular fungi, algae and cyanobacteria. The animals that burrowed beneath these organisms, and presumably fed upon them, are unknown. Early arthropods have been suggested, but as yet no fossil remains of such have been recovered.

By late Silurian times, the situation had become quite different. The first fossils of intact vascular plants have been recorded from these rocks, most of them belonging to the 'Rhyniophyta' (recently considered to consist of several individual groups). Genera such as *Cooksonia* had erect stems with sporangia at the tips, but no leaves. All the records of these plants come from marine sediments, and they may have grown on the edge of a marine lagoon, in an analogue of the present-day saltmarsh environment, or perhaps on small islands (Edwards, 1980; Edwards & Fanning, 1985). Although most of the Silurian plants were not very tall, they had upright growth forms, and presumably cast shade and increased the humidity near the ground, as well as providing a source of detritus in the shape of fallen stems. In other words, they modified the micro-climates near the ground to a considerable degree, and, as will be described in Chapter 2, these microclimates are critical for the survival of animals. Despite these effects, we have records of only one major group of animals from this time. These are fossils that were probably myriapods. Unfortunately, all the specimens from the Silurian are badly preserved, and somewhat ambiguous (Rolfe, 1980, 1985). Besides these, scorpions have been found in rocks of the same age, but these were aquatic, utilizing gills protected by abdominal plates (Briggs, 1987). In summary, then, we can only say that Silurian lands had very poorly developed ecosystems, with very low diversity. Plants and animals were probably limited to fringes around fresh and salt water, and as far as we know the uplands were devoid of macrofauna and macroflora.

## 1.2    Development of terrestrial ecosystems in the Devonian

At the start of the Devonian, several major land masses were situated near the equator, and, although they were beginning to converge, they were still separated from one another and from the southern continent of Gondwanaland which lay over and around the south pole (Fig. 1.3). By the end of the Devonian, most of these land masses had probably converged to form a large continent called Pangaea (Livermore, Smith & Briden, 1985), although it is not certain whether the gap between the northern and southern continents had been closed by then (Scotese *et al.* 1985). It is not known whether this configuration of the land masses itself provided particularly favourable conditions for the development of land plants, or whether changes in current and in climate were influential, but some speculations can be made. Because the lands that would later form North America and north-western Europe were near the equator, and were separated, water currents were able to circulate round them. There was probably little latitudinal variation in climate, and this climate was probably relatively mild. In this climatic regime, during a stage called the Frasnian, the species diversity of marine invertebrates living on the world's continental shelves reached very high levels (Valentine & Moores, 1976), possibly aided by the gradual convergence of the continents, which produced extensive low-lying and coastal areas. It seems likely that this development in the seas was connected with the rapid establishment of terrestrial fauna and flora. While the continents were coming together, but had not yet fused, extensive areas of continental shelf with divergent characteristics, and possessing faunas that had been separated for a very long time, were juxtaposed. Large shallow areas would have provided ideal opportunities for air breathing to develop, both in marine and in fresh waters. At the same time plants with aerial leaves would have experienced ideal conditions. At the end of the Frasnian stage, in the last stage of the Devonian, the Famennian, there was a mass extinction of species

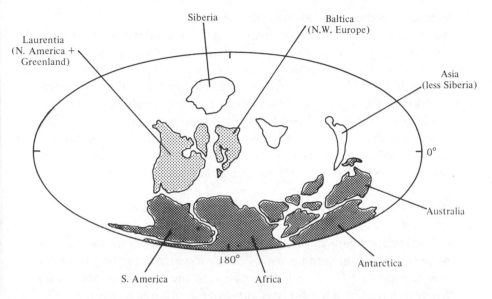

Fig. 1.3. Reconstruction of the distribution of major land masses in the early Devonian. The Old Red Continent (light stipple – Laurentia and Baltica) may or may not have been joined to Gondwanaland (dark stipple – S. America, Africa, Antarctica, Australia). After Scotese, Van der Voo & Barrett (1985) and Livermore, Smith & Briden (1985).

(McGhee, 1989). This may have been due to a global drop in temperature, and was less prominent in fresh water than in marine systems.

Whether or not the above scenario is correct, the Old Red Sandstone continent of the Devonian (essentially the present-day North America plus north-west Europe) saw the subsequent development of extensive terrestrial vegetation and of several terrestrial animal phyla. The Devonian flora was cosmopolitan (Chaloner & Lacey, 1973) and contained, as well as the 'Rhyniophyta', present in the Silurian, Bryophyta (liverworts), Lycophyta (club-mosses), Sphenophyta (horsetails) and Pteridophyta (ferns). Even the early rhyniophytes of the Devonian, such as *Zosterophyllum* and *Psilophyton*, were relatively tall compared with *Cooksonia*, and the later lycophytes and pteridophytes reached the size of small trees (Edwards, 1980). The evolution of this dramatically increased plant cover must have changed terrestrial micro-climates from those of the late Silurian to a marked degree. By providing continuous shade, lessening the effect of wind and retaining a humid atmosphere, by supplying organic detritus and a direct food supply of living material, the plants effectively created cryptozoic niches that allowed the evolution of small, desiccation-intolerant animals. Dominant among these were myriapods and early hexapods, all presumably feeding upon detritus. Centipedes and probably scorpions were present, as the first known terrestrial predators, although the first definitely terrestrial scorpions were recorded from the Carboniferous (Briggs, 1987). Into this environment the early amphibians emerged, probably as small forms dependent upon humid niches, preying upon the varied arthropod fauna. The whole Devonian ecosystem probably depended closely upon the use of cryptozoic habitats, and as far as is known there was no development of a fauna outside them.

## 1.3     The Carboniferous and the development of forests

It is not certain whether the ocean gap between the equatorial Old Red Sandstone continent and the southern continent of Gondwana closed in the late Devonian or in the early Carboniferous, but the closure would have disrupted ocean currents flowing round the continents, and would therefore have disrupted the accompanying moisture-laden winds. The following change of world climate to a more continental one, with latitudinal variation, built up throughout the Carboniferous, and it is possible that the southern part of Pangaea (formerly Gondwanaland) became glaciated. In spite of this latitudinal variation, the climate of the former Old Red Sandstone continent (primarily North America and northwest Europe), which was still near the equator, was tropical, and the luxuriant coal-swamp vegetation provided ideal climatic conditions for the evolution of both arthropods and vertebrates on land.

The coal-swamp forests contained many pteridophytes, but the greatest significant environmental effects were produced by the development of the conifers. These, with methods of reproduction that were somewhat independent of moist conditions, and with tall trunks and a variety of leaf form, produced environmental conditions that were much more constant than those of the Devonian. In these forest conditions the pterygote insects evolved, the land snails appeared and the reptiles evolved from amphibian stock. The ecosystems of Carboniferous time therefore had many parallels with those of today. The vertebrates and insects were the dominant forms on land, living in environments where the conditions were closely controlled by the growth of vegetation. Since vertebrates and insects were the most important herbivores, the direction of evolution of both groups was related to that of plants. The insects also produced carnivorous forms such as the dragonflies, but the role of predator was particularly taken up by the chelicerates, which produced many forms during this period. This early interaction between insects and arachnids has been retained through geological time, and is still one of the dominant characteristics of terrestrial ecosystems.

## 1.4    The Permian and the end of the Palaeozoic

The change of world climate towards a more continental one increased in the Permian, with the consolidation of Pangaea. The Permian period was characterized by the extensive spread of deserts, presumably with extremes of climate, and it is noticeable that during this time no new terrestrial plant or animal groups were recorded. The diversity of marine animals decreased, presumably due to the fusion of continental coastlines, and the obliteration of the large areas of shallow continental shelf.

To summarize the development of terrestrial ecosystems in the Palaeozoic, the most important phase of evolution and diversification probably occurred at a time of widespread mild or tropical climate. The two major groups of terrestrial animals, the insects and the vertebrates, emerged and underwent radiation during these conditions. The present-day distribution of both of these groups suggests that they have retained adaptations that are essentially tropical, and that their world distribution can be envisaged in terms of invasion of the temperate zones from the tropics.

## 1.5    The Mesozoic: Triassic to Cretaceous

The Mesozoic was characterized by the break-up of Pangaea (Dietz & Holden, 1976), and the consequent gradual return of milder conditions and decreasing latitudinal variation in climate. The Triassic saw the evolution of no new groups on land, but in the mild conditions of the Jurassic and Cretaceous the organisms that form the essential parts of present-day ecosystems appeared: the birds and mammals evolved from reptilian ancestors, the various groups of land snails underwent explosive radiations, and the angiosperms came to dominate the vegetation. It is tempting to correlate the environmental influence of the angiosperms with the massive radiations of at least some groups, particularly the land snails. Until recently, it was thought that many land snail groups appeared in the late Cretaceous, but it is now known that both the land pulmonates and one group of land prosobranchs, the Neritacea, were present in the Carboniferous (Solem, 1985). Although Solem has argued that the Carboniferous fossils indicate a very early diversification of pulmonates on land, the records of land snails do not become abundant until the Cretaceous, and it is only then that some of the prosobranch groups such as the Littorinacea and Architaenioglossa appeared on land. While there is no direct evidence to suggest why the terrestrial molluscs should have greatly expanded in diversity and numbers in the Cretaceous, their dependence upon a plentiful calcium supply may be involved. In recent ecosystems the detritus formed in coniferous forests is usually an acid humus because the trees absorb few nutrients from the soil, and leaching produces a resultant downward movement of soluble ions. There are few invertebrates in this type of habitat except for insects and mites. It is only in broadleaved forests that the trees absorb nutrients from deeper layers of soil and effectively return them to the soil surface by the annual fall of leaves. In this type of soil many invertebrates are found, including those with a skeleton or shell of calcium carbonate. If these generalities apply to past ecosystems, the calcium supply in the humus of pre-Cretaceous gymnosperm forests was always poor, and this effectively reduced their rate of colonization by molluscs. With the development of angiosperms, the molluscs came to take a place similar to that found today. It is possible that somewhat similar considerations apply to the origins of the terrestrial decapod Crustacea, but as far as is known these did not arrive on land until the Tertiary.

The general features of the terrestrial insect fauna reached great similarities to that of today in the Jurassic (Hughes & Smart, 1967), except for the Lepidoptera. Lepidopterans are primarily nectar feeders, and they arose in conjunction with the flowering plants in the Cretaceous. The co-evolution of insects and plants is one of the major factors governing present-day ecosystems, and will be treated in more detail in Chapter 10.

### 1.6    Conclusions

Two great phases in which life on land developed have been highlighted in the above account. The first was in the Devonian and the Carboniferous, when terrestrial ecosystems first expanded to cover wide areas. The second was in the Cretaceous, when angiosperms began to supplant the gymnosperms, and the foundations of modern ecosystems were laid. Both these phases occurred when world climate was generally mild, with no great latitudinal variation, and in both the greatest diversification appears to have occurred near the equator. We have no method of accurately reconstructing environmental conditions at these times, but a consideration of animals in transition from water to land in today's tropics may provide some insight, and some later chapters will concentrate on the biology of animals in such tropical habitats as mangrove swamps.

Although the fossil record described above provides an overall view of the origin and development of terrestrial ecosystems, it gives no insight into the mechanisms by which terrestrial animals evolved and became adapted to new environments. In order to understand such mechanisms, it is necessary to visualize the problems facing potential invaders of the land, and this is done in Chapter 2. Chapter 3 then considers the types of evidence available to us in investigating the invasion of land: the fossil record is one of these, but several other approaches can be used with confidence. These introductory chapters provide sufficient background to move on, in Part II, to a consideration of the routes on to land followed by various animal lines. It will be seen that, while the most widespread view is that animals moved on to land via fresh water, many invertebrate groups took more direct routes. In Part III, the adaptations of animals to life on land are reviewed, and the structure of present-day terrestrial ecosystems is discussed.

# 2

## Problems: terrestrial versus aquatic habitats

On chalky and sandy soils, and in the hot villages about
London, the thermometer has often been observed to mount as
high as 83 or 84; but with us, in this hilly and woody district, I
have hardly ever seen it exceed 80; nor does it often arrive at
that pitch. The reason, I conclude, is, that our dense clayey
soil, so much shaded by trees, is not so easily heated through
as those above-mentioned: and besides, our mountains cause
currents of air and breezes; and the vast evaporation from our
woodlands tempers and moderates our heats.
Gilbert White (1788) *The Natural History of Selborne.*

For animal lines moving from water to land, the changes in
physical and chemical characteristics of the environment are immense.
These changes affect all possible life processes, from respiration and
excretion to methods of movement, the functioning of the sense organs and
reproductive mechanisms. The transition must therefore have been made
very gradually, and some of the transition stages are repeated by present-
day groups. Marine Crustacea such as the fiddler crabs, *Uca* spp., for
instance, can be seen to spend periods of time in the air, and in high
temperatures, but at intervals they retreat to their water-filled burrows in
the mud. This allows them to cool down and to re-hydrate. Movement back
and forth across the water/air interface is thus most important for animals
in transition between water and air, yet crossing this interface involves
many problems. The first section of this chapter compares the physical
properties of air and water, and the problems that these create for animals.
Later sections of the chapter consider how these physical properties
combine to make the aquatic and terrestrial habitats so different from each
other. In particular, they deal with the sudden but small-scale changes
within terrestrial habitats, and the development of what are called micro-
habitats or micro-climates.

## 2.1  Physical characteristics of water and air

Some of the physical properties of air and water are shown in Table 2.1. Water is about 1000 times as dense as air, and this means that it provides far more physical support. The density difference is reflected in the quite different skeletal mechanisms of many aquatic invertebrates and their terrestrial relatives, and probably contributes to the inability of some aquatic groups to move on to land. Most of the marine invertebrates that are not arthropods have some form of hydrostatic skeleton (see Trueman, 1975), in which the contraction of opposing sets of muscles acts upon an internal fluid-filled chamber, often a coelom, to produce changes in body shape. Elongation is usually produced by the contraction of circular muscles, and recovery of the original shape by contraction of longitudinal muscles. This mechanism can also function on land, but, because the external surface of terrestrial animals receives little support from the air, the muscles of the body wall must be much thicker than those of aquatic animals. Even then, many terrestrial animals with a hydrostatic skeleton are confined to habitats such as the soil, where they can move through the spaces in the soil matrix. A good comparison between locomotion in marine and terrestrial animals is provided by the marine lugworm *Arenicola* and the earthworm *Lumbricus* (Seymour, 1970). The locomotion of the terrestrial *Lumbricus* is more variable than that of *Arenicola* because it has a more complete metameric segmentation, and pressure changes are to a great extent limited to a few segments at a time. *Lumbricus* also has more inherent rigidity of the body wall, caused by thick muscular layers, and by the additional support of the muscles within the intersegmental septa. It is therefore better adapted both for burrowing into crevices in the soil, and for maintaining its body shape in air. A more extreme example of how aquatic animals depend upon water for support is given by coelenterates such as sea anemones, which utilize the sea water contained within the coelenteron as the hydrostatic skeleton. Without the surrounding sea water to renew this fluid, sea anemones have no basis for muscular antagonism.

Although water provides more support than air because of its higher density, its higher viscosity tends to limit the speed at which animals can move (Gray, 1968). The fastest swimming fish such as the tunny can achieve speeds of just over $10 \, \text{m.s}^{-1}$, and possibly up to $20 \, \text{m.s}^{-1}$, but birds such as swifts can fly at about $40 \, \text{m.s}^{-1}$. Even this comparison is not as straightforward as it seems, because fish can maintain high speeds for only very short times, whereas birds can continue fast flight for long periods. The high speeds achieved by tunny are shown only in bursts, and after

Table 2.1 *Some physical properties of water and air*

|  | Water | Air |
| --- | --- | --- |
| Density (g.ml$^{-1}$ at 20°C, 76 cmHg) | 1.000 | 0.0012 |
| Viscosity (centipoises at 20°C) | 1.00 | 0.02 |
| Oxygen content (ml/100 ml at 20°C) | 0.66 | 20.95 |
| Carbon dioxide content (ml/100 ml at 20°C) | 0.03 | 0.033 |
| Thermal capacity (cal ml$^{-1}$°C$^{-1}$ at 20°C) | 1.00 | 0.0003 |
| Refractive index (at 20°C) | 1.33 | 1.00 |
| Velocity of sound (m.s$^{-1}$ at 20°C, 76 cmHg) | 1486 | 343 |

10–20 seconds they drop to much lower speeds. Similarly, slower fish such as trout can produce bursts of speed up to 1.6 m.s$^{-1}$, but can maintain only 0.6 m.s$^{-1}$ in sustained swimming. Birds such as racing pigeons, on the other hand, can maintain speeds of nearly 20 m.s$^{-1}$ for hours.

These rates of movement are governed not just by the characteristics of the medium in contact with the external surface of the body, but by the ways in which the medium limits the supply of oxygen to the tissues, and the removal of carbon dioxide from them. Air and water have quite different capacities to supply oxygen and carbon dioxide. Because of the higher viscosity of water, it is more difficult to ventilate respiratory organs in water than in air. In addition, while the partial pressure of oxygen may be the same in water as in air, the maximum amount of oxygen per unit volume is 30 times greater in air. The solubility of carbon dioxide in water, in contrast, is greater than that of oxygen by a factor of 25. In consequence, the design of many respiratory systems in terrestrial animals is quite different from those living in water, as will be discussed in Chapter 9.

One other major physical difference between air and water must be emphasized. The thermal capacity of air is more than three orders of magnitude lower than that of water, so that air temperatures may fluctuate rapidly over short time periods or over short distances, while water temperatures are relatively stable. The changes in air temperature cause changes in other factors such as rate of evaporation of water, so that they make the aerial environment much more variable in many ways. The whole topic of variability must therefore be discussed in much more detail (Sections 2.3 and 2.4).

## 2.2     Salt and water supply in water and on the land
Various properties concerning the availability of water and salts are very different in water and on land. Salts are usually readily available in

water, either in high concentrations as in the sea, or in smaller and more variable concentrations as in estuaries and fresh water. On land, in contrast, salts are often not available in solution, and the supply of elements for the building of skeletal material, such as calcium, is dependent upon the rock type, or more particularly upon the soil found above the bedrock. Although, for instance, soils on limestone usually have a high pH and high calcium content, the calcium may be leached out in some conditions so that an acid heath results. In this type of habitat, shelled molluscs are rare, whereas they are common on lime-rich soil.

Most aquatic animals are subject to the osmotic inflow of water, so that water is only a problem insofar as it must be excreted. Some aquatic species find it difficult to obtain water, however, because of differences in osmotic pressure between the external water and their body fluids. The marine teleosts provide a classic example. Because they maintain the osmotic pressure of their body fluids at about one third of sea water, there is a tendency for water to move out of, and not into their bodies. To obtain a water supply, they drink sea water, but then have the additional problem of excreting the extra salts accumulated (Rankin & Davenport, 1981). On land, in contrast, water loss is the normal problem because the water vapour pressure is usually lower than saturation, and is often very variable. Even on land, though, excess water may cause a problem, especially for small arthropods: if their spiracles are blocked, they may drown. Water supply is therefore a major controlling factor for terrestrial animals, and many of the topics discussed in this book are related to it. In particular, the life-styles of terrestrial animals can be more closely related to water supply than to any other environmental variable. Table 2.2 gives a guide to the various life-styles shown by animals found on land.

This classification of the relationships of terrestrial and semi-terrestrial animals to their environment is not meant to be used too rigidly, but it allows at least some attempt to be made at defining which animals are truly terrestrial and which are not. If animals are effectively covered by a layer of water, then they are living as aquatic animals. If they are not so covered, which often means just that they are bigger, as earthworms are usually bigger than soil-dwelling nematodes, then they can be said to be truly terrestrial. This definition results in such forms as soil-dwelling nematodes, mites, rotifers and tardigrades being called aquatic, while burrowing amphibians and earthworms are called terrestrial. Such a division at least has the merit that it fits the known facts concerning terrestrial adaptations in soil dwellers. Thus the nematodes have no obvious adaptations for terrestrial life, whereas the earthworms can reproduce out of water and

Table 2.2 *A classification of life-styles on land*

*1 Aquatic animals*
Those active only when covered by water or a water film; e.g. soil protists and
nematodes.

*2 Cryptozoic animals*
Those requiring constant high humidity, and intolerant of desiccation, usually
found in humid micro-habitats such as leaf litter; e.g. nemertines, many woodlice.

*3 Hygrophilic animals*
Those requiring high humidity for activity, but tolerant of desiccation, often in a
state of aestivation; e.g. snails.

*4 Xerophilic animals*
Those active in dry conditions; e.g. vertebrates, insects, arachnids.

obtain their oxygen from the air; although this is not to say that many of the
problems faced by earthworms are not faced by freshwater animals as well.

Terrestrial animals have a variety of relationships to water supply. Those
that require constant humidity and may be difficult to separate from
aquatic forms are called cryptozoic because they maintain themselves in
humid, often dark environments. Leaf litter and soil are major sites for
cryptozoic animals. Other animals need a water supply for activity, but can
tolerate some degree of desiccation. They pass times of reduced water
supply in some form of suspended animation – either aestivation, or a more
extreme form called cryptobiosis, which is discussed in Section 9.1.6. These
animals are termed hygrophilic. Finally, there are animals that have
reduced their requirements for humid micro-climates, and can be active for
long periods in the general macro-climate. These are the xerophilic
animals, which show the greatest structural and physiological adaptations
to life on land. These adaptations are discussed further in Chapter 9.

**2.3    Aquatic habitats: characteristics and variability**
Because of the high thermal capacity of water, the temperature of
the sea and of lakes remains relatively constant. For instance, in the North
Atlantic Ocean, the average surface temperature varies by only about 8 °C
over the year, and near the equator the average annual variation is only
0.5 °C (Sverdrup, *et al.*, 1942). In the short term, diurnal variations in
surface waters may be as much as 2 °C in the tropics, but deeper water has a
very constant temperature indeed, changing only a few degrees over the
year. Freshwater lakes show both more marked thermal stratification, and
greater temperature variation than the oceans, principally because of their

smaller water mass. Even in temperate lakes, annual surface variations of 20°C are common, and there are often substantial vertical differences at any one time: with deep lakes, there may be a 20°C difference across the thermocline (Hutchinson, 1975).

Generally speaking, then, the smaller the water body, the larger is the temperature range experienced, and in habitats on the margins of water bodies the situation is very different from that in the main water mass. These marginal environments are the regions that must be crossed by aquatic animal lines evolving on to land, and hence this book will be concerned with many of them in detail. For example, the sublittoral marine environment merges into rocky, sandy or muddy shores, saltmarshes and mangrove swamps, and these in turn border on forests, grasslands or deserts. Some of these transitions are obviously more gradual than others, and provide more gentle gradients in physical factors, so that they may form easier routes for invasion by animals. In Part II, some of these environments are considered in more detail as possible routes on to land (Chapters 5–9).

For the moment, it is sufficient to take one example of the variability found in marginal aquatic habitats. A familiar one is provided by marine intertidal pools, common on both rocky shores and saltmarshes. For rockpools, some of the physical and chemical variables have been investigated by Daniel & Boyden (1975). Examples of their measurements of temperature and oxygen levels over 24 h in summer, in a series of pools on the coast of Wales, are shown in Fig. 2.1. Temperatures rose to a maximum of 24°C, and fell to a minimum of 14°C. Oxygen concentration rose above 20 mg.l$^{-1}$ (more than 300% saturation) in some pools, but fell at night to less than 1 mg.l$^{-1}$ (about 3% saturation). Concomitant changes in carbon dioxide concentration and pH occurred: photosynthesis during the day reduced the carbon dioxide concentration in most pools from over 100 mg.l$^{-1}$, in some cases to half this value; while pH rose from night time values around 7.5 to a daytime maximum of 9.5. Such violent changes in the physical and chemical regime undoubtedly restrict the diversity of fauna and flora found within these habitats.

In saltmarsh pools, temperature regimes have been investigated by Marsden (1976). Pools on a marsh in south-east England showed very similar diurnal temperature changes to those in the rockpools discussed above: a minimum of 14°C was recorded in summer, and a maximum of 24°C. In winter, diurnal changes were small – from a minimum of −0.5°C to a maximum of 5°C. Animals in these pools therefore have the double problem of acclimating to long-term changes and tolerating sudden short-

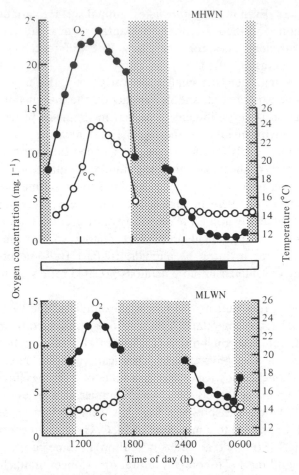

Fig. 2.1. Diurnal changes in oxygen concentration and temperature during summer in two intertidal rock pools in south-west Wales. The high-level pool was near mean high water of neap tides (MHWN) and was exposed to the sun. The lower pool was near mean low water of neap tides (MLWN) and was shaded. Filled circles show oxygen. Open circles show temperature. Stipple shows coverage by the tide. The horizontal bar shows day (open bar) and night (filled bar). After Daniel & Boyden (1975).

term changes. The isopod *Sphaeroma rugicauda*, common in these pools, is well adapted to this fluctuating regime, and is able to alter its activity patterns rapidly in response to temperature change. As will be seen later, the development of tolerance to changes in physical factors shown by marine intertidal animals has in many cases pre-adapted them for life on land. This whole topic is taken further in Chapter 4.

**2.4      Variability on land and the characteristics of terrestrial micro-climates**

The classification of animal responses to water on land given in Table 2.2. suggests a wide variety of different niches within the terrestrial environment. Each of these has its own set of physical characteristics, termed a micro-climate, and in most cases these differ considerably from what is termed the macro-climate – the climate measured by meteorologists. Meteorological readings are usually taken from within screened shelters about 2 m above the ground, so that these readings are relatively independent of the nature and state of the ground. They allow overall comparisons between different latitudes and different environments, because all the readings are taken under comparable conditions. For most animals, however, this macro-climate is not particularly relevant, with the exception of large vertebrates and flying animals. Most animals, after all, live either on the surface of the ground or vegetation, or actually within the soil or the plant tissues. Fortunately, the climate near the ground has now been studied extensively. The following sections are based mainly upon the classic work of Geiger (1965), or upon the excellent modern summary by Oke (1978). Micro-climates have been considered with special reference to insects by Willmer (1982).

*2.4.1      The micro-climate of soils*

When the ground surface is bare of vegetation, the harshest climate is undoubtedly found on its very surface. Here, at the interface between earth and air, solar energy arrives during the day faster than it can be transmitted down into the soil, or re-radiated. The soil surface therefore rises in temperature faster than the air above it or the soil below. At night time, heat is re-radiated from the surface faster than it can be replaced from the soil, and the surface temperature falls more rapidly than that of the air or of the soil at depth. The soil surface therefore experiences extremes of temperature. The phenomenon is well shown by observations made in soils where there is no vegetation to moderate the surface climate (Fig. 2.2).

The temperature regime of the soil below the surface is controlled by, among other factors, the soil's thermal conductivity. If this value is high, heat can move rapidly through the soil, and the extremes at the surface are less. Thermal conductivity of dry sandy soil is about 0.3 watts.$m^{-1}°C^{-1}$, while that of dry peat soil is 0.06 watts.$m^{-1}°C^{-1}$, so that peat surfaces heat up more rapidly than those of sand. To complicate matters, however, the actual temperature changes that occur in the soil also depend upon its

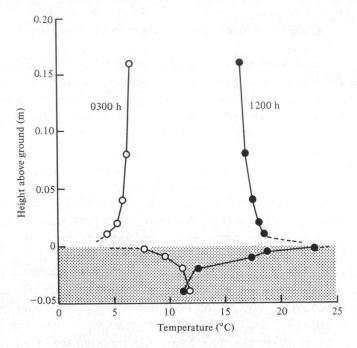

Fig. 2.2. Two temperature profiles near the surface of unvegetated ground in north-east U.S.A. The left profile (open circles) was taken at 0300 h, and shows an inversion type. The right profile (solid circles) was taken at 1200 h, and shows a lapse type. Maximum temperature range is at the soil surface. After Geiger (1965).

thermal capacity – the amount of heat needed to increase the temperature of a unit volume by a unit amount. As pointed out in Section 2.1, the thermal capacity of water is very much higher than that of air, so, if water replaces some of the air spaces within the soil, the total thermal capacity rises markedly. For example, when a dry sandy soil is saturated with water, its thermal capacity more than doubles, while in a peat soil it increases by about seven times. Neither thermal conductivity nor thermal capacity, therefore, provide a true measure of how a soil responds to thermal input or loss, and a more useful term is thermal diffusivity (Oke, 1978). This measures the time required for temperature changes to travel within the soil. Some values are given in Table 2.3. From this table it can be seen that the lowest values of thermal diffusivity are found in peat soils, while clay and sandy soils show higher values which vary depending upon water content. The sand and clay soils, therefore, allow heat to penetrate rapidly, so that, while subsurface soils heat up fairly quickly, the soil surface does

Table 2.3 *Thermal diffusivity of soils*

|  | $m^2.s^{-1} \times 10^{-6}$ |
|---|---|
| Sandy soil – dry | 0.24 |
| – saturated | 0.74 |
| Clay soil – dry | 0.18 |
| – saturated | 0.51 |
| Peat soil – dry | 0.10 |
| – saturated | 0.12 |
| Water (4°C, still) | 0.14 |
| Air (10°C, still) | 20.50 |

*Source:* Oke, 1978.

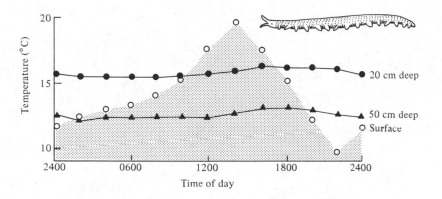

Fig. 2.3. Temperature fluctuations over 24 h in the habitat of the Brazilian onychophoran *Peripatopsis acacioi*, during winter. This species burrows in the soil to depths as great as 1 m, thereby avoiding surface variations in temperature. Open circles and stipple, surface values. Filled circles, 20 cm deep in soil. Triangles, 50 cm deep in soil. After Lavallard *et al.* (1975).

not experience such extremes. Peat soils, on the other hand, allow only slow penetration of heat, and the soil surface may experience violent temperature changes.

These differences between the thermal properties of various soils obviously have enormous consequences for animals living in them, or, more especially, on their surfaces. It will be seen later that many animals use temperature cues to trigger their activity, and to ensure that they stay in favourable micro-habitats. In the meantime, it is appropriate to give some examples of temperature regimes at the soil surface and below it. Fig. 2.3

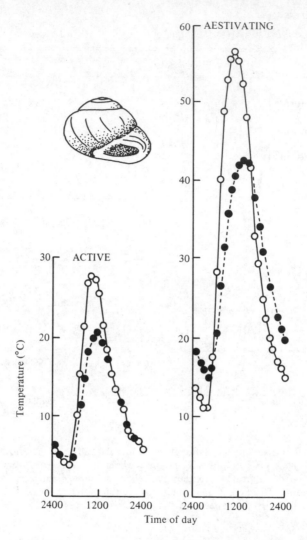

Fig. 2.4. Temperatures in the habitat of the snail *Sphincterochila prohetarum*, which lives in the Negev desert. The left graphs show conditions in March, just after rain had fallen. Snails were actively feeding during this period. The right graphs show conditions in June, when all snails were aestivating under stones. Open circles show temperatures on the exposed desert surface. Filled circles show temperatures under stones. After Steinberger *et al*. (1983).

shows the temperatures in the habitat of a Brazilian species of onycho-phoran, *Peripatopsis acacioi*. This lives in the soil on the sites of former forests. In the summer, the surface temperature seldom falls as low as 17°C, and the animals are found in the top few cm. In winter, when surface temperatures may fall to 10°C, the animals avoid changes in temperature by migrating downwards, often as far as 1 m below the surface (Lavallard *et al.*, 1975). Fig. 2.4 shows temperature fluctuations in a much more extreme environment, the Negev Desert. Here the snail *Sphincterochila proph-etarum* is active when the soil is moist after rains, even when temperatures in the day reach 25°C (Steinberger *et al.*, 1983). When the rains cease, however, temperatures rise to nearly 40°C and the soil dries out. The snails then move under large stones and aestivate.

This last example leads on to the point that, apart from temperature, the other major variable in determining the characteristics of soil habitats is the soil's capacity for water retention. Open textured, wet soils, such as wet sands, have a very low soil moisture tension – in other words, it is easy to extract water from a sandy soil, and the soil tends to dry out rapidly. Clays, at the other extreme, have a high soil moisture tension, and do not dry out so easily. The values of soil moisture tension are perhaps most important when considering the availability of water for plants, since plants absorb most of their water through the roots. There are, however, some animals that extract water from soils by physical suction, and for these the soil moisture tension is critical. The mechanisms used by these animals will be considered later (9.1.1).

### 2.4.2 The micro-climate of the air near the ground

The thermal regime in the air immediately above the ground was briefly discussed above. Once again, extremes are greater nearest to the soil surface. The thermal regime of the air is also to some extent dictated by the thermal properties of the soil. For instance, the dry peat soils discussed in Section 2.4.1 absorb solar energy but retain it at the soil surface, which therefore becomes very hot. A 'lapse' profile develops above it (as at 1200 h in Fig. 2.2), in which temperature declines with vertical height. At night, however, the peat soil radiates heat, but heat flow to the soil surface from subsurface layers is slow, so the surface becomes very cold. An 'inversion' profile then develops above it (as at 0300 h in Fig. 2.2), in which air temperature *increases* with vertical height. In contrast to this situation over peat soils, the climate above wet sand or clay is much less extreme because heat transfer occurs rapidly in these soils. Besides this, the incorporation of water in the soils means that heat is lost by evaporation

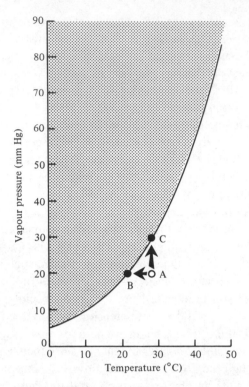

Fig. 2.5. The variation of saturation vapour pressure of the air with temperature. Saturated conditions are shown stippled. Point A shows a saturation deficit of 10 mm Hg. If temperature decreases to B, or if vapour pressure rises to C, dew will form. After Edney (1957).

during the day, as well as by radiation, and this helps to reduce temperature extremes.

For animals, the amount of water vapour in the air is in many cases more crucial than the actual temperature values. The maximum amount of water vapour that the air can hold is, however, dependent upon temperature: at low temperatures the air can hold little moisture, but the saturation value rises exponentially as temperature rises. The amount of water vapour that the air can hold is usually known as its saturation vapour pressure. The variation of this with temperature is shown in Fig. 2.5. From this it can be seen that, if any sample of unsaturated air is cooled sufficiently, it will reach its saturation vapour pressure. When it does so, water will condense, and the temperature at which condensation occurs is called the dew-point temperature. In most climates, the supply of water that the dew provides for terrestrial animals is quite small – the majority of water is provided by precipitation in the form of rain or snow; but for some desert animals the falling of dew may be almost the only water source.

The vertical distribution of vapour pressure above the soil is determined mainly by the movement of water across the soil/air interface. As suggested in the above paragraph, the downward precipitation of dew is normally a small fraction of the water supplied to the earth, and in the absence of precipitation the movement of water is mainly upwards from the soil to the air. Highest humidities therefore tend to occur near the soil surface, so that, when the temperature regime there *is* favourable, the surface may be the least stressful region for animal activity: animals will lose least water by evaporation in highest humidities, and, as emphasized already, water supply is perhaps the most crucial physical limitation for terrestrial fauna.

### 2.4.3    The influence of plants on terrestrial micro-climates

Although it is convenient, as an introduction, to discuss the micro-climates immediately above and below the surface of bare ground, the majority of ground is in reality influenced by the growth of plants. One of the primary effects of plant cover is to raise the effective site of energy exchange above the ground surface: the sudden changes in temperature therefore tend to occur near the top of the plants, and not on the soil surface. For cryptozoic soil animals, or those living in leaf litter on the soil surface, vegetation cover therefore greatly reduces the temperature extremes. A good example is shown in Fig. 2.6, in which the micro-climates of a grass and clover meadow are outlined. Here the temperature maximum is not at the soil surface, but 30 cm above it. The effect of vegetation in reducing the amount of radiation that reaches ground level can easily be appreciated when it is noted that the total surface of vegetation growing in a meadow is some 20 to 40 times the area of the ground on which it grows (Geiger, 1965).

Fig. 2.6 also shows the decrease in humidity above the ground surface. In this case it is presented as the 'saturation deficit' or drying power of the air: technically the amount by which the water vapour pressure falls short of saturation. Near the ground, the value is near 0 mm Hg, i.e. the air is nearly saturated. A maximum value for saturation deficit, i.e. the driest air, is reached about 40 cm above ground level. In general, the air within a vegetation stand is more humid than outside, and this needs some further discussion, since a constant high humidity is one of the factors that greatly influences cryptozoic and hygrophilic animals.

On a bare soil surface, the water input consists mainly of precipitation –

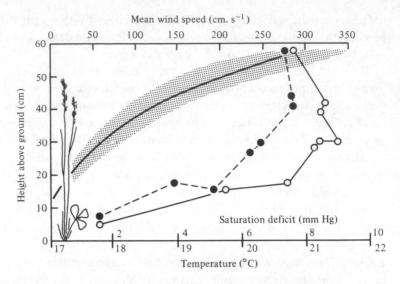

Fig. 2.6. Profiles of temperature, humidity and wind speed in a grass and clover meadow in Scotland during sunshine in June. Open circles show temperature. Filled circles show saturation deficit. Solid line shows mean wind speed and stipple indicates the range of variation. After Waterhouse (1955).

rain or snow – and, to a very slight degree, frost or dew. This input is balanced by movement of the water into the soil, run-off and evaporation. When vegetation cover is present, several notable differences occur in this water balance. Input is still mainly from precipitation, but this is interrupted near the top of the vegetation cover, mainly by leaves. Much of this water is retained by the leaves and does not reach the ground, so that direct precipitation on, say, a forest floor is less than that received by the canopy. The general range of this interception varies from about 10 to 50% of rainfall for different species of tree, and under different circumstances. The arrival of rain on the forest floor is also delayed: only when the leaves are saturated does much water drip from them, either directly to the ground or by running down the stems. Because the degree to which precipitation is intercepted varies so much between plant species, the micro-climates on the floors of forests of different tree species are very diverse.

Turning now to the processes by which water is lost from the soil, the comparison between bare soil and soil with vegetation cover is also striking. Since the rate of input of water to a forest floor is delayed by the canopy, the floor receives much fewer sudden bursts of water input than an unprotected soil. The forest soil can therefore absorb water more effectively, and run-off does not cause such violent soil erosion as in deforested areas. Some of the water from soil surfaces beneath vegetation is removed by evaporation, as it is from bare soils. The wind speed within stands of vegetation tends to be low, however (Fig. 2.6), and the rate of eddy diffusion is therefore also low. Water vapour within a vegetation stand therefore moves out of the stand only very slowly, by diffusion. This is in great contrast to open surfaces, where, although there is a thin boundary layer above the soil, air movements rapidly remove water that has evaporated from the soil. The greatest difference between vegetated and non-vegetated surfaces, however, lies in the importance of the movement of water through the vegetation itself. Water is absorbed by the roots, passes up the stems and is then lost by transpiration from the leaves. This transpiration accounts for the greatest portion of the water lost from the canopy to the atmosphere. It is usually combined with direct evaporative losses, and called evapo-transpiration. Because transpired water is lost from leaves via the stomata, the opening or closing of which can be altered depending upon the plant's requirements, it follows that the total water lost by evapo-transpiration is not purely a process determined by the physical properties of the environment, but is under the control of the plants. Stomata are closed when the plant's water supply declines, so that the water reserves within a vegetation stand are actively conserved by the vegetation itself. The climate within forests or other vegetation stands is thus regulated relatively closely, and animals living within these micro-climates take advantage of the relative constancy.

### 2.4.4 *Interactions between animals and their micro-climates*

Up to this point, terrestrial micro-climates have been described as they are measured by physical instruments, on the assumption that animals live passively in their environment. In fact, however, the introduction of any new matter – animate or not – into a small space will inevitably alter the micro-climate. Furthermore, if the properties of the introduced materials differ from those already in the space, their physical variables such as temperature may not conform to those already present. To take a purely physical example, Parry (1951) used blackened brass discs as models of insects, and investigated the effect of height above the ground on their temperatures, in direct sunshine. The brass discs were always warmer than nearby shaded air. The amount by which they were warmer – their 'temperature excess' – increased as they were placed nearer to the ground, because of the increase in air temperature and the decrease in wind velocity, without change in the net radiation load. This radiation load may itself be greatly affected by the colour, shape and orientation of the body concerned, as was shown by Digby (1955) using measurements of temperature inside recently killed insects. Heat was lost from both the models and the insects by convection as well as by radiation. Since the rate of loss by convection varies inversely with linear dimensions, larger bodies have higher equilibrium temperatures than smaller ones: an important point for, say, arthropods living in desert conditions.

From this example, it is apparent that the physical characteristics of an animal, as well as its position in the micro-climate, affect its body temperature. Another excellent example is provided by desert pulmonate snails during aestivation (Schmidt-Nielsen *et al.*, 1971). Fig. 2.7 shows how temperature is distributed in an aestivating snail on the desert surface. During the dry season, this surface may reach 65°C. When aestivating, *Sphincterochila boissieri* withdraws its tissues from the lower, body whorl, which rests on the ground, so that they are effectively insulated from the ground by an air space. They are also then slightly above the ground surface and will therefore experience slightly lower temperature extremes (see 2.4.1). In addition, the shell of this species, like that of most desert snails, is thick and white, and very reflective. As a consequence, about 95% of the incident radiant energy is reflected. The overall result is that the snail's tissues never reach a temperature higher than 50°C, which is within the tolerance limits for the species.

Many animals do more than rely upon their physical characteristics to generate a temperature excess or to remain cooler than their environment. They regulate their body temperature either by behavioural adjustments, or by physiological processes. Those that use solely behavioural mechanisms are referred to as ectotherms, while those that use physiological mechanisms are called endotherms. It is not appropriate here to go into details of the various homeostatic mechanisms employed by animals. These are dealt with in many text books, particularly with reference to the vertebrates (e.g. Schmidt-Nielsen, 1975; Hardy, 1972). It is essential, however, to understand just how independent of external temperatures some of the invertebrates can be. In particular, it is to some extent this independence that has allowed many arthropods to invade lands with quite extreme climates, and to become the most numerous of terrestrial animals.

Behavioural mechanisms for temperature control are well shown by many spiders and insects (Willmer, 1982). Among the spiders, each species has a 'preferred' temperature, usually achieved by basking in the sun to warm up, and by retreating to the shade to cool. This preferred temperature may change at different times of year. For instance, it changes when the

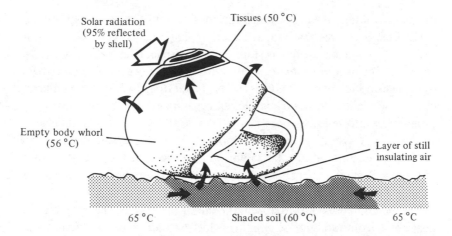

Fig. 2.7. Temperature distribution and heat flow in and around the desert snail *Sphincterochila boissieri*. The thick white shell reflects solar radiation. The tissues of the snail withdraw to the top whorls which are cooler than the body whorl. Solid arrows show the direction of heat flow. After Schmidt-Nielsen, Taylor & Shkolnik (1971).

female is carrying eggs: in *Pirata piraticus*, a spider that lives in *Sphagnum* bogs, it rises from a norm of 20–22 °C to 26–28 °C, and the females migrate to the surface of the moss to 'sun' their eggs. Other spiders do not carry their eggs with them, and have adopted different mechanisms of temperature control. *Theridion saxatile*, for instance, lives in a nest suspended by silk threads. When this becomes too hot from absorption of radiant heat, the female takes the eggs out of the nest and hangs them in the shade underneath, returning them at night (Nørgaard, 1956).

Regulation of the spider's own body temperature can also be carried out by basking in the sun. The burrowing Australian species *Geolycosa godeffroyi*, for instance, emerges from its burrow as external ambient temperature rises, and basks until its body temperature reaches about 40 °C. It can maintain this temperature, with some variation, throughout the day by basking and retreating into its burrow in turn. This behaviour is very similar to the basking behaviour of many reptiles (Avery, 1979).

Many similar examples can be quoted for insects, but desert tenebrionid beetles have been particularly well studied (Henwood, 1975). In the Namib Desert of south-western Africa, two of these beetles provide contrasting thermoregulatory patterns. *Onymacris plana* inhabits sand dunes, and selects dunes facing in appropriate directions at different times of day. In order to absorb solar radiation in the morning, it moves to east-facing dunes. At midday it burrows into the loose sand of dune faces to avoid surface temperatures that would be lethal. Then in the late afternoon, it emerges to move to west-facing slopes to absorb heat from the sun again. In contrast, *Stenocarpa phalangium* lives in the stony gravel plains between dunes. The option of burrowing is therefore not open to it, and at midday it climbs the highest stones available, elevates its extremely long legs, and points the white, reflective posterior part of its abdomen directly at the sun. In this way it avoids the highest temperatures of the desert surface, and minimizes the amount of solar radiation absorbed during the hottest part of the day. A similar effect is seen in another desert tenebrionid, *Onymacris rugatipennis* (Fig. 2.8).

For the animals so far discussed, the major problem is one of balancing heat input from radiation against heat lost by radiation, conduction and convection. For insects in flight, the process of temperature regulation is more complicated, especially because a very great deal of heat is produced by the flight muscles. Unless these are actively cooled, they must, in large insects, heat up to 30°C or more above ambient temperature during continuous flight. They have, therefore, become adapted to operate at these high temperatures, and as a corollary it follows that the muscles must be warmed up if the insects are to be able to initiate flight (Heinrich, 1975). The two main ways in which this is done are basking, as for example in butterflies, and 'shivering' (the use of small-amplitude wing-beats) as in dragonflies and moths. Many insects maintain a fairly constant thoracic temperature once they are flying, and are therefore acting temporarily as homeotherms. The eastern tent caterpillar moth, *Malacosoma americanum*, for example, maintains a thoracic temperature between 34 and 41 °C when the ambient temperature varies from 10 to 30°C. This does not mean, however, that all large flying insects do so. The desert locust *Schistocerca gregaria* does *not* maintain a constant temperature in flight; and the temperature of *small* insects such as *Drosophila* always remains near ambient.

A final example of temperature regulation in insects is that of the African dung beetles, which gather dung into balls before rolling it away prior to consumption. When these beetles are at rest their body temperature is the same as ambient (Bartholomew & Heinrich, 1978). While walking, body temperatures may, or may not, be elevated. Prior to flight, body temperature is greatly elevated, probably by isometric contractions of the flight muscles. Take-off temperature for large beetles is of the order of 40°C, and this temperature is maintained in flight in both thorax and abdomen. During other activities, such as ball-making and ball-rolling,

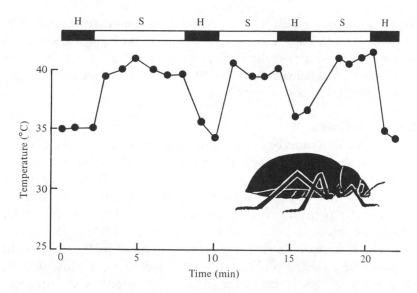

Fig. 2.8. Thoracic temperature in a tenebrionid beetle from the Namib desert, *Onymacris rugatipennis*. The bar shows when the beetle was orientated with its head towards the sun (H) or with the body at right angles to the sun's rays (S). Orientation towards the sun reduced body temperature by about 10°C. After Edney (1971).

body temperatures are nearly as high as those during flight: this allows faster collection of dung in a situation of intense intra- and interspecific competition. In this example, then, the beetles are homeothermic whenever it is advantageous to be so, and not just in flight. The contrast with small, cryptozoic insects, in which temperature varies entirely passively with the environment, could hardly be more extreme.

In birds and mammals, control of body temperature is more complete and finely adjusted than in the arthropods, but the mechanisms used in its control have far-reaching consequences. In particular, the major cooling mechanism employed involves evaporation of water, either as sweat or as respiratory loss during panting. This is a distinct contrast to the situation in insects, where evaporative loss of heat is very small. Use of evaporative cooling means that, unless a copious external water supply is present, the animals run the risk of dehydration. Since areas of high temperature such as deserts are also often very dry, many desert mammals have evolved mechanisms to avoid evaporation. Small animals such as kangaroo rats use burrows in the daytime, while larger forms like camels actively reduce the fine-tuning of their temperature controls, and allow body temperature to rise during the day. Excess heat is then lost by radiation at night (Schmidt-Nielsen, 1964). In general, however, the temperature control mechanisms of birds and mammals allow them extreme independence from the environment, and individual species often have much more wide-ranging distributions than those of poikilotherms.

## 2.5    Conclusions

The differences between aquatic and terrestrial habitats, outlined in this chapter, are so enormous that the likelihood of animal life moving across the water/air interface may seem extremely low. Nevertheless, many animal lines have invaded the land, and this is essentially because the aquatic and terrestrial environments intergrade: it is not necessary for animal lines evolving on to land to switch instantly from the constant conditions in water to the very variable conditions on dry land. Life in the fringes of the aquatic environment, where conditions are quite variable, pre-adapted animals for life on the fringes of the terrestrial environment, where conditions are more variable. The fringing habitats are therefore of special importance when considering the invasion of land, and they will be examined in detail in Part II of this book. In the meantime, it is important to consider just what evidence is available to us in attempting to trace the origins of land animals, and this is the subject of the next chapter.

# 3

## The evidence

but geology alone can tell us nothing of lands which have
entirely disappeared beneath the ocean.
A. R. Wallace (1869) *The Malay Archipelago*, London: Macmillan.

Most of the movements of animals on to land that produced
today's terrestrial communities occurred at some considerable time in the
past, as emphasized in Chapter 1. It might seem, therefore, that palaeontological evidence would provide the most appropriate method to investigate these movements. There are, however, severe limitations and problems
in the use of this evidence, and various other lines of attack have proved to
be very useful. This chapter outlines the four main types of evidence about
the origins of terrestrial animals, and discusses the usefulness of each.

### 3.1    Geological evidence
The discussion in Chapter 1 of the evolution and change of terrestrial
communities from Silurian times to the present depends upon the
collection and interpretation of fossil evidence. Since this is our only *direct*
way of tracing the origins of terrestrial animals, it is extremely important
that the accuracy of such information be understood. There are several
problems with the interpretation of palaeontological data, and three
specific areas of uncertainty are discussed below.

### 3.1.1    *Problems of preservation*
In general, only the hard parts of animals are preserved as fossils.
Obvious examples are the shells of molluscs, and the bones and teeth of
vertebrates. When examined and analysed in detail, these skeletal remains
can be very informative concerning the soft parts of the animals. This,
however, is often the case only when we *also* can examine present-day
relatives with a more-or-less similar structure. For example, it is possible to
reconstruct a fossil bivalve or a brachiopod in some detail because we know
what tissue structures in modern equivalents are reflected by specific

characteristics of the shell. With groups that have died out, this is far from easy. The greatest problem, however, is the lack of preservation of animals that had no hard skeleton. This problem is emphasized by the few examples in which soft-bodied animals *have* been preserved.

The most famous and well-worked example of a fossil soft-bodied fauna is that of the Burgess Shale of the Cambrian (Conway Morris, 1986). These shales contain a remarkable assemblage of marine invertebrates, some with hard skeletons, but many without. The fauna was preserved at the foot of an algal reef by sudden blanketing flows of fine sediment (Fig. 3.1). These produced anaerobic conditions which inhibited the processes of decay, and also presumably prevented access by scavenging animals. Among the invertebrates are polychaete worms, priapulids, sponges and many arthropods, and the soft-bodied animals contribute about 86% of this diverse array. Such a high proportion of soft-bodied fauna causes problems in the interpretation of communities in which soft-bodied animals have *not* been preserved: if these other communities had a similar composition, our interpretations of them are based on only 20% of the original diversity. This means that, even after extensive examination of fossil-bearing strata, we may still have very little insight into the structure of ancient communities, and may make quite wrong assumptions about the animals that played dominant roles in them.

Lack of preservation of animals as fossils is also a real problem for studies of the origin of invertebrate groups. The origins of the hexapods, for example, are obscure partly because of the lack of discovery of any forms prior to the Collembola found in Devonian rocks. The Devonian forms show no great differences from those of today, so that they provide us with no evidence about hexapod ancestry. It is quite likely that hexapod ancestors were not large or heavily armoured, and the likelihood of their preservation – and discovery – is very low. The occasional preservation of later insects in amber provides important evidence about insect diversification, but not about insect origins.

For the vertebrates, although the story of early radiations is fairly clear, here again the problem of origins is confused by lack of preservation. Many authorities agree that the echinoderms are close to the chordate line, but debate continues about the position of the group known as the calcichordates (Jefferies, 1986; Jollie, 1982). This group has been well-preserved in fossil form, so that its relationships can at least be considered from a basis of fact. In contrast to this situation, many zoologists believe that at a later stage in evolution the vertebrates arose by neoteny from the larva of a sessile form, probably a tunicate. Berrill (1955) argued that a tunicate larva

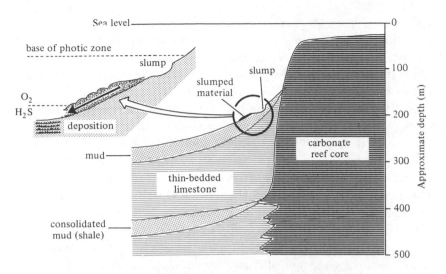

Fig. 3.1. Diagrammatic section to show the conditions in which the Burgess Shale was deposited. Mud was banked up against the edge of an algal reef, and slumps in this mud occurred, burying organisms. In selected sites, especially next to the reef cliff, these deposits became anoxic, and soft-bodied animals were prevented from decaying. The inset shows possible details of the slump regions, known to be near the base of the photic zone because of the presence of algae. After Whittington (1985) and Conway Morris (1986).

became neotenous and failed to metamorphose, thus being able to exploit the rich plankton of the early seas as a food source, and forming the first vertebrate. If such a course actually occurred, there is only a remote chance that these forms would be preserved.

In summary, we can say that while fossil forms can be extremely valuable in interpreting the course of animal evolution, the lack of preservation of small soft-bodied animals often leaves gaps at vital places in the record. Worse than this, the inevitable emphasis on skeletal remains may bias our interpretation of some communities. On the other hand, of course, fossil evidence provides the only direct picture of past communities, and, as long as its shortcomings are recognized, it provides an invaluable tool. With this in mind, it is important to consider two other possible difficulties in interpreting palaeontological evidence.

### 3.1.2 Relationships of fossil forms to modern forms

Even when the soft external structures of animals are well preserved as fossils, it is not always possible to be sure of the relationship of these fossil forms to modern forms. A good example is provided by referring again to the Burgess Shale fauna. One of the animals in this community was *Aysheaia pedunculata* (Whittington, 1978). In external features this bears some resemblance to the modern Onychophora, and it has been postulated that it is a marine forebear of the modern *Peripatus*. It differs in many respects, however. It is not at all clear, for instance, how the cephalic appendages of this animal relate to the antennae of modern Onychophora. These cephalic appendages appear more like limbs, and are accompanied by papillae round the mouth (Robison, 1985; Briggs & Conway Morris, 1986), but it is probable that *neither* of these two sets of structures is homologous with structures in modern forms. In reality, the phylogenetically 'important' characters of Onychophora, such as the mechanism of limb movement, the presence of tracheae and the so-called nephridia, are almost entirely internal, and a study of external features only is bound to be inconclusive. To elucidate the relationships of *Aysheaia* is therefore at the same time of crucial importance, and extremely difficult.

Another example of the difficulty of relating modern to fossil forms is given by the early evolution of the Crustacea. Besides the forms that can definitely be assigned to the Crustacea there are large numbers of fossils sometimes placed in the Pseudocrustacea. These have many crustacean characteristics, but do not fit into any known relationship to present-day crustaceans. A great variety of further arthropods were also abundant in the Burgess Shale together with *Aysheaia*. Examples are *Sidneyia* and *Marrella* which, as discussed by Briggs (1985), show no obvious relationships to any higher group. As with the hexapods, then, it is impossible to reconstruct the origins of many arthropod groups from examination of fossil evidence: the fossils give us some insight into the early radiation of types, but, because of the infrequent preservation of soft-bodied forms, they seldom provide us with the crucial intermediate stages.

### 3.1.3    The environment of fossil forms

It is very important to be able to relate the distribution of fossil species to their environment. While this might seem relatively easy, it can in fact produce quite controversial situations. The fossil record of the early vertebrates shows how complex the problem can be. Until quite recently, the majority of evidence suggested that the first vertebrates evolved in fresh water. More recent conclusions, however (e.g. Bray, 1985; Halstead, 1973), suggest that the earliest vertebrates preserved as fossils in the Lower and Middle Ordovician lived in a marine environment. From these early forms movement into fresh water is thought to have occurred in the Silurian and

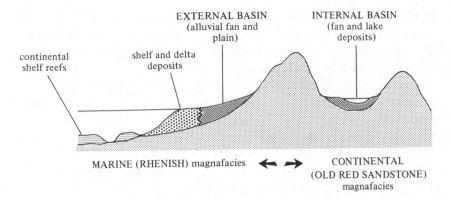

Fig. 3.2. Diagrammatic section to show conditions under which the major deposits were formed in Devonian times. Fan and lake deposits of internal (freshwater) basins closely resemble fan and alluvial deposits of external (marine) basins. Hence it is often difficult to be certain whether early vertebrate remains have a marine or freshwater origin. After Bray (1985).

Devonian, and the groups then radiated into marine, brackish-water and freshwater environments.

From these changes of opinion about the conditions in which the vertebrates evolved, it is evident that the determination of whether sediments and their contained fossils are marine or fresh water in origin is far from straightforward. The conditions in which sediments were deposited can only really be assessed by an analysis of sediment types and of the faunas contained within them. The sediments of the Siluro-Devonian transition among which early vertebrate fossils have been found were probably laid down in a mixture of marine, freshwater and brackish-water conditions (Fig. 3.2). In this complex situation, with some material being washed down by rivers into the sea, and preserved alongside material of marine origin, it is not perhaps surprising that confusion and disagreement about conditions have arisen. In other cases, the origin of the sediments may be assigned by analysis of the fauna contained within them; but again this is not always satisfactory because the further back in time the sediments were laid down, the less is known about the habitat preferences of the animals preserved in them.

A second example of the difficulty of defining the conditions under which past species lived is provided by the gastropod molluscs (Solem & Yochelson, 1979; Solem, 1985). Several genera of land snails have been described from beds in the late Carboniferous. A few of these occur on upright stems of ferns, suggesting that this was their natural habitat. Most, however, have been found in freshwater deposits so that their original habitat can only be surmised. The mixtures of worn and fresh shells, and their size-groupings, suggest that some species at least were washed out of leaf litter in much the same circumstances as happens today in areas of high seasonal rainfall. Many litter species live today in tropical areas that are not swamps, but are flooded seasonally, and these are often preserved in freshwater deposits similar to those of the Carboniferous. It must be admitted, however, that the information about the ecology of fossil land snails derived from such comparisons is very speculative.

The various types of evidence that relate to the environments of fossil organisms have been discussed at a recent symposium (Trans. R. Soc. Edinb. 1989). There are many indicators from which suggestions can be made about the characteristics of past environments. Temperatures, for example, can be inferred from isotope ratios and the molecular ratios of certain organic compounds, although absolute values are still subject to debate. Atmospheric composition can be inferred from other isotope ratios. An overall estimate of the habitat of a fossil assemblage can be gained from a combination of techniques, as described by Briggs & Clarkson (1989) for early Carboniferous crustaceans. This involves the use of geological evidence *plus* comparisons with present-day faunas, and such an approach seems to be the most useful one.

## 3.2    Evidence from present-day distribution

Together with the palaeontological evidence, where this is available, the present-day distribution of various groups suggests likely routes of invasion of land. For example, a terrestrial group with no freshwater relatives either at the present day or in the fossil record may be suggested to have moved on to land directly from the sea. In contrast, a group with close freshwater relatives may well have evolved from these and not directly from marine forms. Examples of this type of evidence are discussed in Chapter 4. By itself, evidence like this cannot be conclusive, but it may lead at least to the erection of a hypothesis about origins. This hypothesis can then be tested by looking for further evidence that may support or destroy it. In particular, the evidence from comparative physiology and anatomy provides a fruitful line of enquiry

## 3.3    Evidence from comparative anatomy and physiology

As will be discussed in Chapter 9, nearly every facet of life has been altered with a change from water to land. Such changes, by which various species are adapted to particular environments, have been investigated by comparative physiologists. A good example of such an approach is that of Edney (1977), who has worked out many of the adaptations of terrestrial arthropods. The information gained by such comparative physiologists can also be used to suggest possible mechanisms by which animal lines have, in the past, been able to move on to land from aquatic habitats (Little, 1989b). This procedure is bound to be a dangerous one because of the flexibility of physiological mechanisms, and the rapidity with which they can change to adapt animals to the environment. Nevertheless, there are some cases in which physiological processes inherited by terrestrial animals from marine ancestors differ greatly from those inherited from freshwater ancestors. To use the characteristics of such physiological processes as evidence of ancestry must always be rather speculative, and the approach should probably never be carried to the limit by assuming that mechanisms present in living species evolved directly from those of other living species. It can, however, indicate general trends that may have occurred in evolution, and its use is discussed in detail for the processes of salt and water balance in Chapter 4.

Because the evidence derived from comparative studies constitutes by far the greatest weight of information available, it is worth while briefly summarizing the possible lines of attack at this point. These lines are used in Chapters 4 to 8, and the actual adaptations of animals to life on land are summarized in Chapter 9.

### 3.3.1 Salt and water balance

Differences in environment appear to have led many times to the evolution of mechanisms for conserving water. These may be physiological, structural, behavioural or, more often, a combination of all of these. At the same time, most terrestrial animals probably have specific mechanisms for obtaining a salt supply, although little is known of these. Overall, the effectiveness of these two sets of mechanisms is reflected in the composition of the body fluids. A comparison of the composition of the body fluids of terrestrial animals with that of their aquatic relatives may often, therefore, suggest whether the terrestrial forms are more closely related to marine or freshwater ancestors. This in turn may indicate whether a particular group has invaded the land from fresh water or the sea. This topic is enlarged in Chapter 4.

### 3.3.2 Respiratory mechanisms

Mechanisms for obtaining oxygen and for eliminating carbon dioxide that are efficient in water often do not function well in air. For example, gills tend to collapse in air, causing a change from a very large surface area for gas exchange to a very small one. Most terrestrial animals, therefore, have evolved lungs or vascularized air-filled cavities instead of gills. The ways in which this has occurred, and the details of the respiratory structure and function may give a guide to ancestry.

### 3.3.3    *Nitrogenous excretion*

In water, many animals produce toxic nitrogenous wastes such as ammonia. Because ammonia is very soluble, it can be carried away in the large volumes of water used for ventilating the respiratory organs. On land, where there is no such water supply, different mechanisms have been evolved, although in many cases these still involve excretion of ammonia. The relation of excretory product to habitat therefore constitutes a possible line of evidence, although as will be seen this relation is far from simple.

### 3.3.4    *Reproduction*

The processes leading to fertilization and the development of eggs vary immensely in marine, freshwater and terrestrial animals. In some cases larval forms occur. In others there is direct development, and there may be some form of care of the young. In many environments internal fertilization takes the place of the broadcasting of sperm and eggs. Use of such characteristics to determine ancestry must be used with caution. Nevertheless, the details of reproductive methods often reflect the origins of a group, and within specific lines it is sometimes possible to trace factors that are related to origins in specific environments. In particular, the use of large or small eggs, the production of different types of spermatophores and the development of brooding will be considered in later chapters.

### 3.3.5    *Movement*

Mechanisms of movement are closely related to the physical properties of the environment, as emphasized in Chapter 2. Because such physical properties of sea water and fresh water as density and viscosity are rather similar, it is unlikely that any really different methods of movement would be related to an origin or radiation within these two habitats. The evolution of particular gaits, and the ways in which animals have developed rapid methods of movement have their own fascination, however, and are discussed in Chapter 9.

### 3.3.6    *Sense organs and behaviour*

Together with an increase in speed of movement on land there may have been an increase in the complexity of behaviour patterns in terrestrial, as opposed to aquatic animals. Even this conclusion may be doubted, however, because there have been relatively few detailed investigations of the behaviour of aquatic animals. If behaviour *is* in general more complex on land, it might imply that some terrestrial sense organs are more sophisticated than those of aquatic animals, and this line of approach may be useful. As with movement, however, there seems to be little basic difference in the sense organs of animals derived from ancestors of different environmental origins.

## 3.4    Negative evidence

The last line of evidence about the factors that have allowed the invasion of land derives from discussion of the types of marine animals that have *no* terrestrial representatives. This 'negative' evidence must, however, be treated cautiously, as it is only too easy to suggest that a particular aquatic type could never have become terrestrial because of a particular structural form or physiological adaptation, only to find that there are other animals on land with very similar features. Nevertheless, it is worth while discussing the reasons that lie behind the absence of such groups as the echinoderms and the squids on land, as these may provide clues about the barriers at the water/air interface.

## 3.5    Conclusions

Of the four sources of evidence about the origins of terrestrial animals, it is evident that the geological evidence has considerable drawbacks. It may give us a record of ancestral types, and in some groups such as the vertebrates it is possible to trace a series of fossil species that become more and more independent of water. For most groups, however, the fossil record is so fragmented that, as emphasized by Rolfe (1985), we have to depend almost entirely upon evidence from distribution, and comparative anatomy and physiology. In Part II of this book, this approach will be taken. By examining the environments intermediate between water and land, and the adaptations of the animals in these intermediate habitats, a picture of how terrestrial ecosystems evolved will be presented.

# PART II

**Routes on to land**

Following on from the general introduction given in Part I, Part II considers the importance of various evolutionary routes from water to land. First, Chapter 4 analyses the origins of several animal groups, using evidence from mechanisms of water balance to establish whether they moved from sea to land directly, or whether they first invaded fresh water. The following chapters then deal with more specific routes. Chapters 5, 6 and 7 cover the direct routes from the sea: Chapter 5 is concerned with mudflats and sands, while Chapter 6 covers marine areas with dense vegetation – saltmarshes and mangrove swamps – and Chapter 7 covers rocky and shingle shores. Finally, Chapter 8 discusses the alternative 'indirect' route through estuaries and fresh water.

# 4

## The origins of terrestrial animals in relation to salt and water balance

> Nowhere more than in the area swept by the rise and fall of the tide do more varied, difficult or changing conditions exist. Hence it is not surprising that various dwellers of the sea in crossing this barrier have become changed to land creatures.
> G. C. Klingel (1959) *Wonders of Inagua*, London: Robert Hale Ltd.

### 4.1    The routes on to land, and evidence about them

When considering the origins of terrestrial organisms, one of the fundamental topics to discuss is that of the routes taken by various evolutionary lines, and the consequences these may have had for their later representatives. Many textbooks have given prominence to the evolution of terrestrial vertebrates, with the likelihood of their early movement into fresh water and the subsequent development of terrestrial forms from aquatic species adapted to stagnant swampy conditions. This emphasis on the vertebrate story has tended to obscure the possibility that invertebrate lines adapted to marine conditions may have invaded land directly from the sea, with no intermediate freshwater stage. Nevertheless, there is now much evidence to suggest that the majority of present-day terrestrial invertebrates belong to stocks which evolved directly from marine ancestors, with many consequences for the mechanisms by which they are adapted to terrestrial life. In this chapter an introduction to the alternatives of marine and freshwater origins will be given, and the ways in which these origins have affected subsequent mechanisms of water balance will be discussed.

### 4.2    Distributional evidence about origins

There are three main sources of evidence about the ways in which animal lines have produced terrestrial groups, as outlined in the Introduction: the evidence from fossils, evidence from present-day distribution, and evidence from comparative anatomy and physiology. Combining the first two categories, it is possible to get some idea of which terrestrial groups have, or have in the past had, freshwater relatives, and which groups have

Table 4.1 *Distribution of some invertebrate taxa containing terrestrial species*

|  | Sea | Fresh water | Land |
|---|---|---|---|
| *Annelida* | | | |
| Polychaeta | X | — | X |
| Oligochaeta | X | X | X |
| Hirudinea | X | X | X |
| *Mollusca, Prosobranchia* | | | |
| Neritacea | X | X | X |
| Architaenioglossa | — | X | X |
| Littorinacea | X | — | X |
| *Mollusca, Pulmonata* | | | |
| Basommatophora | X | X$^t$ | X |
| Stylommatophora | — | X$^t$ | X |
| *Crustacea, Isopoda* | | | |
| Oniscidea | X* | — | X |
| *Crustacea, Amphipoda* | | | |
| Talitroidea | X | X* | X |
| *Crustacea, Decapoda* | | | |
| Astacura | — | X | X |
| Anomura | X | — | X |
| Brachyura | X | X | X |
| *Chelicerata* | X | X$^t$ | X |
| *Onychophora* | — | — | X |
| *Myriapoda* | — | — | X |
| *Hexapoda* | X$^t$ | X$^t$ | X |

*Notes:* X* indicates supralittoral species only.
X$^t$ indicates species thought to be secondarily adapted from terrestrial lines.

only marine relatives. This distributional evidence is summarized for some groups in Table 4.1, which is essentially an expansion of Table 1.1.

From this table, it may be suggested that for certain terrestrial lines we have no evidence of the existence of any freshwater taxa – notably for the majority of polychaete annelids, littorinacean gastropods and anomuran decapods. To this list may be added further groups for which the freshwater taxa are known to have arisen secondarily from the land (most of the hexapods, pulmonate gastropods and chelicerates), or for which they are not on the direct line of evolution of the terrestrial groups (amphipods, isopods). Other lines, however, have freshwater representatives, and it is not possible from distributional evidence alone to make hypotheses about the origins of the terrestrial platyhelminthes, nemertines, oligochaetes, hirudineans, neritacean and architaenioglossan prosobranchs, brachyurans and vertebrates.

## 4.3    Physiological evidence about origins

At this stage it is profitable to investigate the possibility of using evidence from comparative anatomy and physiology, both to reinforce the suggestions made from distributional evidence, and to erect hypotheses for the origin of those terrestrial groups with freshwater relatives. It may be thought initially that physiological characters would be so highly labile and adaptive that they would not be useful as records of an organism's evolutionary history. A consideration of several groups has, however, shown how significant the physiological evidence can be, as discussed by Little (1989b).

In order to establish the credibility of such evidence, some examples will be discussed in detail. These will centre particularly on osmoregulation and kidney function, because one of the prime adaptations necessary for movement on to land is the regulation of an organism's water balance. It must be emphasized that many other physiological systems have been adapted in the move from aquatic to terrestrial habitats – mechanisms of nitrogen excretion, movement, respiration, reproduction and the sense organs, to name some of the major categories. The evolutionary pressures on some of these aspects will be discussed in subsequent chapters, and the mechanisms involved will be brought together in Chapter 9. Here, however, we begin with the adaptations of salt and water balance found in the prosobranch gastropods, as an example of the use of physiological evidence.

## 4.4    Marine vs freshwater routes: prosobranch gastropods

Several groups of prosobranch gastropods have terrestrial species. In particular, two superfamilies provide an interesting comparison. One of these, the Littorinacea, contains marine and terrestrial snails. The other, the Architaenioglossa, contains freshwater, amphibious and terrestrial snails.

### 4.4.1    The Littorinacea

This superfamily contains four families, but only two are considered here. The Littorinidae comprises marine forms common worldwide on rocky and soft shores, and in the tropics in mangrove swamps. The Pomatiasidae contains terrestrial forms common in the warmer parts of the Americas, in Africa and in the Mediterranean countries, but also found further afield with a few species in India, and in northern Europe, where *Pomatias elegans* is common on calcareous soil. Although many of the

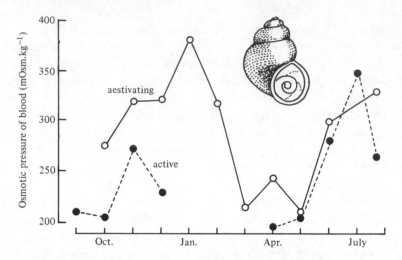

Fig. 4.1. Variation in the osmotic pressure of the blood of *Pomatias elegans* kept in terraria at 25°C. Open circles show aestivating snails. Closed circles show active snails. After Rumsey (1972).

pomatiasids are found in forest leaf litter, they are often common on trees and on limestone cliffs, and so are exposed to extremes of desiccation.

Many of the marine species (Littorinidae) are very tolerant of a wide range of salinities, but none of them have been shown to osmoregulate: the osmotic pressure of the blood is always slightly higher than that of the external medium, but they are unable to regulate this osmotic pressure in the face of changing salinities. Not surprisingly, then, the kidney has been shown to produce urine with the same osmotic pressure as the blood, and with similar ionic composition (Rumsey, 1973). That is to say, the kidney is not involved in any control of the inorganic composition of the blood. Broadly speaking, the animals therefore survive by tolerating osmotic changes rather than by any methods of regulation.

This picture is mirrored by members of the terrestrial family (Pomatiasidae) which are very tolerant of desiccation (Fig. 4.1) – and therefore of rising osmotic pressure of the body fluids – and in which the kidney does not appear to be involved in osmotic or ionic regulation. The urine has the same osmotic pressure and ionic composition as the blood, although there may be some slight re-absorption of ions in the mantle cavity (Rumsey, 1972). It is possible that glands opening on the sole of the foot may be involved either in producing a hyper-osmotic secretion, or in absorbing salts and water from the soil. If this is so, these glands, and not the kidneys, are the effective osmoregulating organs.

Table 4.2 *Ionic composition of the blood of marine and terrestrial species of the Littorinacea*

| Species | Osmotic pressure (mOsm.kg$^{-1}$) | Na$^+$ | K$^+$ | Ca$^{2+}$ | Mg$^{2+}$ | Cl$^-$ | SO$_4$$^{2-}$ | HCO$_3$$^-$ |
|---|---|---|---|---|---|---|---|---|
| | | (% cations) | | | | (% anions) | | |
| **Marine** | | | | | | | | |
| Littorina littorea | 1098 | 87.9 | 3.0 | 2.3 | 11.2 | 95.9 | 2.8 | 1.0 |
| **Terrestrial** | | | | | | | | |
| Pomatias elegans | 254 | 79.7 | 4.5 | 11.8 | 1.6 | 81.1 | 3.5 | 8.7 |
| Tropidophora ligata | 267 | 81.0 | 3.3 | 6.8 | 2.1 | 78.0 | 2.0 | — |

*Source:* Rumsey, 1972.

The inorganic composition of the blood of pomatiasids is very similar to that of the marine littorinids (Table 4.2). Added to this, the osmotic pressure of the blood of the pomatiasids is remarkably high for terrestrial gastropods – about 25% of the osmotic pressure found in the marine littorinids. As will be seen when considering some other terrestrial gastropods, the osmotic pressure of the blood can be less than 10% of that of marine species.

Within the Littorinacea, the terrestrial forms therefore show several similarities with their marine relatives, in terms of the physiology of salt and water balance. If we follow the hypothesis erected from distributional evidence, this physiological evidence can be said to support the idea that terrestrial Pomatiasidae could have evolved directly from marine ancestors.

### 4.4.2   The Architaenioglossa

This superfamily presents an entirely different picture. Here there are three component families (or more, depending on various taxonomic views), of which two will be considered: the freshwater Viviparidae and the terrestrial Cyclophoridae (sometimes itself regarded as a superfamily, and split into a number of families). The Viviparidae contains snails common in temperate regions, usually found in slow-running waters with muddy substrates. The Cyclophoridae contains snails of a variety of shell form, most common in the tropical parts of Indonesia and the Americas. None of the cyclophorids is conspicuous, and, although some are arboreal and others are found in rocky habitats, the majority are confined to forest leaf

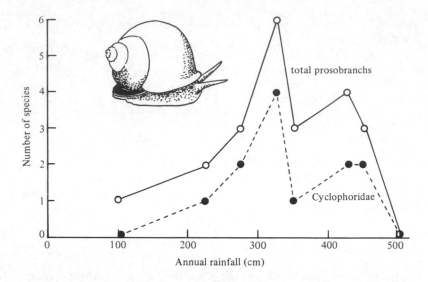

Fig. 4.2. Maximum number of species of terrestrial prosobranchs collected at sites in Papua New Guinea, in relation to annual rainfall. High numbers occur only where rainfall is more than 200 cm.yr$^{-1}$, but at very high rainfalls numbers decline, possibly because of leaching from the soil. Open circles show total species of prosobranchs. Filled circles show Cyclophoridae. After Andrews & Little, 1982.

litter. Many are found in areas of high rainfall (Fig. 4.2), and are therefore not for the most part exposed to desiccating conditions, but during dry periods many probably aestivate.

Of the freshwater species, the physiology of water balance has been investigated in most detail in the European *Viviparus viviparus*. In osmotic terms, it is a typical freshwater invertebrate, unable to withstand salt concentrations greater than about 10% sea water (Little, 1965). The osmotic pressure of its blood is only 68 mOsm.kg$^{-1}$ in fresh water, or about 7% of the osmotic pressure of sea water. In concentrations up to 10% sea water, it is an osmoregulator, maintaining the osmotic pressure of its blood higher than that of the outside medium. In this case, the organs responsible for osmoregulation are the kidney and the pallial ureter, which actively re-absorb salts from the initial ultrafiltrate to produce a hypo-osmotic urine. The contrast with the marine littorinids is striking.

The terrestrial cyclophorids are remarkably similar in salt and water balance to the viviparids just described. The osmotic pressure of their blood is low – about 7 to 8% of sea water – and therefore very similar to that of

*Viviparus*. The composition of the blood is regulated by the kidney, which produces a copious flow of hypo-osmotic urine. The cyclophorids are in osmotic terms so similar to the viviparids that they have been called 'aquatic snails living on land' (Andrews & Little, 1972).

The physiological evidence within the Architaenioglossa therefore strongly supports the idea of a close evolutionary relationship between the freshwater and terrestrial families, and the idea that the terrestrial taxa evolved not directly from marine ancestors but from a freshwater line.

### 4.4.3    *Littorinacea and Architaenioglossa: selective pressures affecting water balance*

The physiological contrast between terrestrial pomatiasids (probably with a direct marine origin) and terrestrial cyclophorids (probably with a freshwater origin) is extreme. Pomatiasids do not use the kidney for salt and water balance, but are good at tolerating changes in osmotic pressure produced by desiccation. Cyclophorids, on the other hand, produce hypo-osmotic urine at a high rate, but cannot tolerate high osmotic pressures. Pomatiasids have blood with a high osmotic pressure, while cyclophorids have a low osmotic pressure. Both are found in leaf litter, but pomatiasids, despite their lack of osmoregulatory ability, are also found in desiccating conditions, while cyclophorids, with extensive osmoregulatory ability, cannot tolerate desiccation. This apparently contradictory state of affairs may reflect the selective pressures acting on ancestral forms in particular environments.

In the marine intertidal zone, relatively few invertebrates are osmoregulators. Osmoregulation in this range of salt concentrations is expensive in terms of energy (Potts, 1954), and among invertebrates the strategy of tolerance is more widespread than that of regulation. There is, in fact, little selective pressure on marine shores to produce hypo-osmotic urine, because the effect of desiccation is to increase water loss, not to increase the intake of water. Marine invertebrate lines moving on to land have in general inherited the ability to tolerate wide changes in osmotic pressure. They are often able to colonize relatively desiccating environments because of their ability as tolerators.

In fresh water, in contrast, all invertebrates are osmoregulators. The majority produce hypo-osmotic urine, which eliminates water moving into the animals osmotically. There is, however, little selective pressure in fresh water to tolerate changes in osmotic pressure: the composition of fresh water is relatively constant, and the regulatory abilities take care of small-scale changes. Freshwater lines moving on to land have often inherited the ability to osmoregulate, and to produce a hypo-osmotic urine. In very humid environments, such as forest leaf litter, this allows them to excrete excess water; but linked with the inherited ability to osmoregulate, they have also inherited the inability to withstand wide changes in internal osmotic pressure. This in turn has limited them to very humid habitats on land.

## 4.5    Blood composition as a guide to animal origins

From the comparison of the mechanisms of salt and water balance involved in two different superfamilies of gastropod, and from the general discussion above, it can be argued that some lines of physiological evidence are useful in helping to assess the evolutionary origins of present-day terrestrial groups. The two most likely physiological characters are now discussed.

### 4.5.1    Osmotic pressure

The most widely available physiological character reflecting the ancestry of animal groups is probably the osmotic pressure of the blood. This character is the end-product of all the osmoregulating mechanisms in the animal. As discussed for two groups of snails, above, many animals with a relatively high osmotic pressure probably have a direct marine origin, while those with a low osmotic pressure have probably descended from freshwater lines. It will be pointed out later that there are several well-known exceptions to this generalization. On the whole, however, the osmotic pressure of the blood reflects that of the ancestral type because, as Potts (1985) has remarked: 'Once on land the blood concentration is no longer subject to strong selection.'

Using this character, together with the evidence from distribution, it is now possible to speculate upon the probable routes on to land used by a variety of animal groups. To give some background for this speculation, Table 4.3 shows some figures for the osmotic pressure of the blood of a variety of terrestrial animals. These are average figures, and the spread is in fact much wider than shown. Nevertheless, it can be concluded that the lowest values are found in the prosobranch gastropods derived from freshwater ancestors, while higher values are found in gastropods that have been derived from marine ancestors, and in crustaceans. The correlation between osmotic pressure and ancestry is a good one. It is not a rigid condition, however, for the reason that the osmotic pressure of the blood is governed by a number of factors, and is therefore not strictly related either to the characteristics of an animal's environment, or to the environment of

Table 4.3 *The osmotic pressure of the blood of some terrestrial animals*

| Species | Taxonomic position | Osmotic pressure ($mOsm.kg^{-1}$) | Possible route to land |
|---|---|---|---|
| **Nemertea** | | | |
| *Argonemertes dendyi* | Hoplonemertea, Prosorhochmidae | 145 | marine littoral |
| **Annelida** | | | |
| *Lumbricus terrestris* | Oligochaeta, Lumbricidae | 165 | fresh water |
| **Mollusca** | | | |
| *Poteria lineata* | Prosobranchia, Cyclophoridae | 74 | fresh water |
| *Pomatias elegans* | Prosobranchia, Pomatiasidae | 254 | marine littoral |
| *Helix pomatia* | Pulmonata, Helicidae | 183 | saltmarshes |
| *Agriolimax reticulatus* | Pulmonata, Limacidae | 345 | saltmarshes |
| **Crustacea** | | | |
| *Arcitalitrus dorrieni* | Amphipoda, Talitridae | 400 | marine littoral |
| *Porcellio scaber* | Isopoda, Porcellionidae | 700 | marine littoral |
| *Holthuisana transversa* | Decapoda, Sundathelphusidae | 517 | fresh water |
| *Cardisoma armatum* | Decapoda, Gecarcinidae | 744 | marine littoral |
| **Vertebrata** | | | |
| *Bufo bufo* | Amphibia, Anura | 205 | fresh water |
| *Bufo americanus* | Amphibia, Anura | 310 | fresh water |

*Source:* Little, 1983; data from various authors.

its ancestors. For example, while it is true that most marine invertebrates have blood with a high osmotic pressure, it is not true that the blood of all freshwater invertebrates has a *low* osmotic pressure. Specific examples will be discussed in later sections, after some consideration of the relevance of the ionic composition of the blood as evidence for ancestry.

### 4.5.2  Ionic composition

The osmotic pressure of the blood of most animals can be accounted for almost entirely by its inorganic constituents, and it is of interest to discuss the relevance of this inorganic composition to ancestry. It is appropriate to begin with Macallum's (1926) hypothesis that when marine organisms first appeared, sea water had an ionic composition far different from that of today, and that many animals have retained ions in their body fluids in proportion to this early composition, long after the composition of the sea had changed. Macallum's idea now seems unlikely on two grounds: one, that early sea water was probably not very different from that of today (Mackenzie, 1975); and two, that the blood of most marine invertebrates follows closely any changes in the composition of sea water. Burton (1973), for example, reviewed the factors regulating the composition of both body fluids and tissues. However, a variant of Macallum's hypothesis has been raised by Spaargaren (1978). He used the common properties of the blood of present-day marine animals to calculate the probable composition of ancient sea water. Various other authors have considered the significance of the ratios of specific ions and groups of ions in the blood. One of these which received early attention is the ratio of the concentration of monovalent cations ($Na^+$ and $K^+$) to divalent cations ($Ca^{2+}$ and $Mg^{2+}$). Lutz (1969) concluded that marine invertebrates have a ratio of approximately 10, while in freshwater invertebrates it is approximately 20. He also suggested that in marine invertebrates chloride concentration is greater than sodium concentration, whereas in freshwater invertebrates the reverse is true. It is useful here to discuss these suggestions, and to see how terrestrial invertebrates fit into the general picture. Most relevant analyses are available for molluscs and crustaceans, and discussion will concentrate upon these two groups.

Some figures for blood composition in marine, freshwater and terrestrial gastropod molluscs and crustaceans are given in Table 4.4. Marine gastropods have a monovalent: divalent cation ratio of just over 7, which reflects their almost total lack of ionic regulation: the ratio for sea water itself is 7.5. In freshwater gastropods the ratio is little different, although in fresh water itself the ratio is much lower. This reflects the fact that, while sodium and potassium are reduced to about 10% of their concentration in marine forms, calcium is retained at 20% of the marine concentration, while magnesium has been reduced to only 5% or less of the concentration in marine forms. Calcium levels are probably maintained in equilibrium with the calcium carbonate of the shell. This trend in freshwater gastropods is continued in the terrestrial species, where similar values for calcium are found in species with a freshwater ancestry, such as species of *Poteria*. *Pomatias elegans*, the terrestrial species with direct marine ancestry discussed in Section 4.4.1, has a calcium concentration even higher than that of marine species. This introduces another point, namely that some of the calcium present is probably bound to proteins. Without the *activities* of each ion species, instead of overall concentrations, it is not likely that a coherent picture will emerge. At present, then, it appears that the monovalent: divalent cation ratio betrays little of species ancestry in gastropods. The same may be said for the chloride: sodium ratio. This is

Table 4.4 *Ionic composition of the blood of selected molluscs and crustaceans*

| Species | $Na^+$ | $K^+$ | $Ca^{2+}$ | $Mg^{2+}$ | $\dfrac{Na^+ + K^+}{Ca^{2+} + Mg^{2+}}$ |
|---|---|---|---|---|---|
| **Mollusca, Gastropoda** | | | | | |
| Marine | | | | | |
|   *Strombus gigas* | 503 | 11.1 | 11.2 | 60.2 | 7.2 |
| Fresh water | | | | | |
|   *Viviparus viviparus* | 34 | 1.2 | 5.7 | <0.5 | 5.9 |
| Terrestrial | | | | | |
|   *Poteria lineata* | 31 | 1.8 | 5.1 | 1.5 | 5.0 |
|   *Pomatias elegans* | 110 | 6.0 | 16.5 | 2.5 | 6.1 |
| **Crustacea, Isopoda** | | | | | |
| Marine | | | | | |
|   *Ligia oceanica* | 586 | 14 | 36 | 21 | 10.5 |
| Fresh water | | | | | |
|   *Asellus aquaticus* | 137 | 7.4 | — | — | ? |
| Terrestrial | | | | | |
|   *Porcellio scaber* | 227 | 7.7 | 14.7 | 10.9 | 9.2 |
| **Crustacea, Decapoda** | | | | | |
| Marine | | | | | |
|   *Pachygrapsus crassipes* | 465 | 12.1 | 11.4 | 29.2 | 11.8 |
|   *Carcinus maenas* | 525 | 12.7 | 14.3 | 21.2 | 15.2 |
| Fresh water | | | | | |
|   *Paratelphusa hydrodromous* | 330 | 6.8 | 7.8 | 7.8 | 21.6 |
| Terrestrial | | | | | |
|   *Sudanonautes africanus* | 207 | 6.0 | 11.8 | 10.6 | 9.5 |
|   *Holthuisana transversa* | 270 | 6.4 | 15.7 | 4.7 | 13.6 |
|   *Gecarcinus lateralis* | 468 | 12 | 17.3 | 7.6 | 19.3 |
| Sea water | 475 | 10.1 | 10.3 | 54.2 | 7.5 |
| Fresh water (soft) | 0.3 | 0.01 | 0.1 | 0.04 | 2.3 |
| Fresh water (hard) | 2.2 | 1.5 | 4.0 | 1.7 | 0.7 |

*Note:* Values mM.1⁻.
*Source:* Little, 1983; data from various authors.

linked with the concentration of bicarbonate in the blood, because bicarbonate makes up a large part of the anion fraction in freshwater and terrestrial gastropods, and is in fact relatively constant (Rumsey, 1972). Bicarbonate is involved in equilibrium with the shell, together with calcium. In summary, although the osmotic pressure of gastropod blood to some extent reflects ancestry, use of the ionic ratios to imply ancestral environments (as in 4.4.1) must be adopted with great caution. Ionic composition of marine and freshwater ancestors may not have imposed limitations on terrestrial forms, which have therefore altered compositions fairly freely.

In the Crustacea the same ratios show a great scatter of values. The semi-terrestrial crab *Sudanonautes africanus* has a monovalent: divalent cation ratio of about 10, and Lutz (1969) has claimed that this suggests a marine origin because it is similar to the ratio in marine crabs. However, the variation in the ratio in terrestrial crabs is enormous. In *Gecarcinus lateralis* it is 19.3, and this species definitely has a direct marine origin, shown by its annual migrations back to the sea to spawn. Two particular problems must be pointed out here. One, the proportion of cations bound on proteins, has already been mentioned. The second is the very variable degree of reduction of magnesium found in the blood of marine crustaceans. In both the marine crabs shown in Table 4.4, magnesium concentrations are about half those in sea water. However, concentrations vary from about 10% to over 90% of the concentration in sea water, in the whole range of marine Crustacea so far studied (Prosser, 1973). This large variation is partly responsible for the variation in divalent cation total, and therefore for the variation in monovalent: divalent cation ratio. In general the large variations suggest that individual ion levels are so closely tied to physiological requirements that they can change radically when crustaceans colonize new environments. In contrast, the general level of osmotic pressure tends to be retained from aquatic to terrestrial habitats.

## 4.6    Marine vs freshwater routes: some problematical groups

In this section and in Sections 4.7, 4.8 and 4.9, the routes taken on to land by a variety of animal groups will be discussed in the light of what is known about their mechanisms of water balance. This discussion will not cover all groups, but will consider some representative lines for which there is a reasonable amount of evidence. Of the soft-bodied invertebrates, some evidence for the oligochaetes and nemertines will be considered here, but general discussion of the interstitial forms, including nematodes, will be reserved for Chapter 5. The flatworms will not be dealt with in detail. Terrestrial planarians are probably most closely related to freshwater forms, at least as evidenced by the similarity of their excretory systems (the protonephridia), and an indirect origin *via* fresh water is therefore suggested.

Of the arthropod groups, the chelicerates will not be dealt with in detail because they have no primitive present-day forms in an intermediate position between water and land. All chelicerates possess haemolymph with a high osmotic pressure (360–480 mOsm.kg$^{-1}$; Little, 1983), which would suggest a direct marine origin. Their origins have been considered by Selden & Jeram (1989), who have concluded from fossil evidence that it is possible that *some* scorpions moved on to land from fresh water.

The origins of the Onychophora and Myriapoda, on the other hand, are not at all clear. Onychophora and diplopod Myriapoda have osmotic pressures from 150–236 mOsm.kg$^{-1}$, while chilopod Myriapoda have osmotic pressures of 318–498 mOsm.kg$^{-1}$ (Little, 1983). Neither possesses other facets of salt and water balance that point clearly to specific routes on to land, and, to add to this confusion, their origins were probably associated, at some stage, with those of the insects. They are considered further in Section 4.9.3 and in Chapter 5.

## 4.7    Marine vs freshwater routes: Crustaceans

All terrestrial crustaceans have body fluids with high osmotic pressure, but, because of the wide diversity of osmoregulatory mechanisms found in aquatic crustaceans, this may represent either a marine or a freshwater origin. It is appropriate to consider the isopods first, and then to contrast them with the amphipods and decapods in subsequent sections.

### 4.7.1    Isopods

There is no evidence that terrestrial lines of isopods evolved in fresh water, and both the physiological and distributional evidence points to a direct marine origin. Most present-day suborders of the order Isopoda are marine, and all the terrestrial forms are contained within one suborder, the Oniscidea. This is probably a polyphyletic grouping, with separate Infra-orders having independent origins from marine ancestors (Vandel, 1965; Holdich, Lincoln & Ellis, 1984). The Infra-orders each have modern marine and terrestrial species, but here we will take as an example the Ligiamorpha. The most well-known marine species in this Infra-order is *Ligia oceanica*, which lives in the rocky marine intertidal at high levels, often in crevices or in shingle. On land, *Armadillidium vulgare* is common in grassland on calcareous soils, and is therefore subject to desiccation. Species of *Ligidium* occur in very damp habitats such as marshes and river banks, and other species occupy many diverse terrestrial habitats: *Porcellio* spp. live in garden soil and woodlands, *Platyarthrus* spp. live in ants' nests, and *Armadillo* spp., *Venezillo* spp. and *Hemilepistus* spp. live in deserts.

The osmoregulatory ability of several species of *Ligia* has been examined (Wilson, 1970), and most show wide osmotic tolerance coupled with some capacity for osmoregulation. Some species can also maintain their

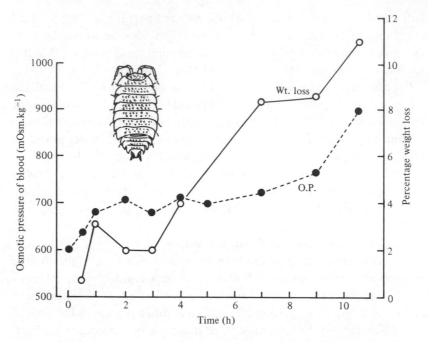

Fig. 4.3. Changes in weight and the osmotic pressure of the blood of the isopod *Porcellio scaber* during desiccation. Osmotic pressure remained relatively constant during a phase of rapid weight loss, suggesting a regulatory process. Open circles, weight loss. Filled circles, osmotic pressure. After Horowitz (1970).

haemolymph hypo-osmotic to that of the external medium at high external concentrations. This ability to hyporegulate is most important for animals that are exposed to desiccation, including the truly terrestrial forms, and maintenance of a constant osmotic pressure in the face of desiccation has been shown for *Porcellio* and *Armadillidium* (Horowitz, 1970). When *Porcellio scaber* was exposed to desiccation, the weight diminished in a linear fashion; but the osmotic pressure, after rising slightly, remained constant for many hours before rising steeply (Fig. 4.3). Similar results have also been obtained for *Armadillidium vulgare* (Price & Holdich, 1980), showing that in the species tolerant of dry conditions this ability is not uncommon. In species that inhabit damper conditions, such as *Oniscus asellus*, no such ability exists. We know little of the details of this osmoregulatory process, but it could be closely related to that involved in the hypo-osmotic regulation of the marine *Ligia*. A review of water balance in terrestrial species was given by Warburg (1987).

Table 4.5 *Ionic composition of the blood of isopod crustaceans*

| Species | Osmotic pressure (mOsm.kg$^{-1}$) | Na$^+$ | K$^+$ | Ca$^{2+}$ | Mg$^{2+}$ | Cl$^-$ |
| --- | --- | --- | --- | --- | --- | --- |
| | | | | (nM.l$^{-1}$) | | |
| Marine | | | | | | |
| *Ligia oceanica* | 1160 | 586 | 14 | 36 | 21 | 596 |
| *Ligia italica* | — | 613 | 16.5 | 34.6 | 19 | 704 |
| *Tylos latreilli* | — | 577 | 23.9 | 25.5 | 55 | 636 |
| Fresh water | | | | | | |
| *Asellus aquaticus* | 273 | 137 | 7.4 | — | — | 125 |
| Terrestrial | | | | | | |
| *Oniscus asellus* | 577 | 230 | 8.2 | 16.7 | 9.1 | 236 |
| *Porcellio scaber* | 701 | 227 | 7.7 | 14.7 | 10.9 | 279 |

*Source:* Little, 1983; data from various authors.

The haemolymph of isopods is thus controlled in its composition to some degree, and it is worth comparing analyses for aquatic and terrestrial forms (Table 4.5). The osmotic pressure of haemolymph in the terrestrial species *O. asellus* and *P. scaber* is over half that of the marine *Ligia* spp., and about twice that of the freshwater *Asellus aquaticus*. The situation is therefore very like that with the terrestrial pomatiasid gastropods (Section 4.4.1) which probably evolved directly from marine ancestors. Terrestrial isopods almost certainly evolved directly from marine forms similar to *Ligia*. If, instead, they had originated from freshwater ancestors, we might expect osmotic pressures of around 270 mOsm.kg$^{-1}$, or less. The analyses of ionic composition support this view. Although it has been concluded above (Section 4.5.2) that in general the ionic composition is very labile in evolutionary terms, the composition of haemolymph in *Ligia* spp. is very similar to that of *Oniscus* and *Porcellio*, and quite dissimilar from that of freshwater forms such as *Asellus*. It is easy to derive the composition of these terrestrial forms from that of *Ligia*, since all the measured ions are approximately halved. Simple dilution could therefore explain the composition. This is not so in *A. aquaticus*, however, where the potassium concentration is equal to that in terrestrial forms, while the osmotic pressure is only half. It may be, then, that while the ionic ratios discussed in Section 4.5.2 give no clue to ancestry, the values of specific ionic concentrations *may* be useful in this respect.

To summarize the situation for isopods on land, we may conclude that they have a small degree of direct regulatory ability allowing them to control the composition of their body fluids by physiological means, but that behavioural mechanisms which maintain them in appropriate micro-habitats are very significant. In this they resemble their marine ancestors. From these ancestors they have also inherited the ability to tolerate changes in water content and osmotic pressure, reflecting the fact that the behavioural mechanisms are seldom precise enough in nature to be able to control the composition of body fluids to within narrow limits. This use of *tolerance* instead of *regulation* will be discussed in more detail later.

### 4.7.2    Amphipods

Much less is known about terrestrial amphipods, and their origins, than about isopod lines. Partly, at least, this is because most terrestrial amphipods are found in tropical and south temperate zones, although there are terrestrial species in Japan. All terrestrial forms belong to the family Talitridae, which also contains species living high up on marine beaches (e.g. *Talitrus saltator, Orchestia gammarellus*) and at the edge of freshwater habitats (e.g. *Orchestia cavimana*). The distribution of a Japanese species, *Orchestia platensis*, strongly supports a recent littoral marine origin for some terrestrial lines: within this species, one subspecies *O. platensis platensis* is found on marine beaches among damp seaweed, while another subspecies, *O. platensis japonicus*, is found in the leaf litter of damp evergreen and deciduous forests throughout Japan (Tamura & Koseki, 1974).

The story is more complex than this, however, because there are now thought to be several relatively separate talitrid lines (Bousfield, 1984). Of these, the line containing *Orchestia* spp. is not common on land, but is well adapted to the litter zones on tidal beaches. These 'beachfleas' can osmoregulate well, being able to hyper-osmoregulate at low external concentrations, and to hypo-osmoregulate at high concentrations (Morritt, 1988). A second line, containing the burrowing sandhoppers such as *Talitrus* spp., are rather similar physiologically. Neither group can tolerate a great deal of desiccation, in comparison with, say, the isopods, but most can tolerate wide variations in external salinity.

A third line contains the true landhoppers, of which a good example is *Arcitalitrus dorrieni* – a species native to Australasia, but now also found in Europe. *Arcitalitrus* cannot osmoregulate particularly well, and is unable to hypo-osmoregulate at high external concentrations (Fig. 4.4), but like the other groups it can tolerate a wide range of salinities. Perhaps surprisingly, it is even *less* able to tolerate desiccation than the two groups of beachfleas (Morritt, 1987). This combination of characteristics may

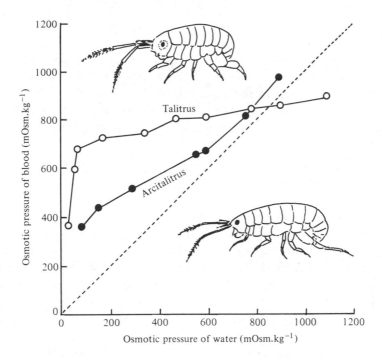

Fig. 4.4. Osmotic regulation in talitrid amphipods in aerated water. The supralittoral species *Talitrus saltator* (open circles) shows good regulation, but the terrestrial species *Arcitalitrus dorrieni* (filled circles) cannot regulate so well. The dotted line shows the line of no regulation. After Morritt (1988).

have come about by movement on to land directly from the high intertidal zone to forest leaf litter: *Arcitalitrus* resembles the primitive high-shore amphipod *Hyale nilssoni* in its osmoregulatory and tolerance patterns. It is particularly instructive to compare *A. dorrieni* with an *Orchestia* species that has invaded the fringes of fresh water – *Orchestia cavimana*. This species has adapted to fresh water by evolving the ability to produce hypoosmotic urine. *Arcitalitrus* produces isosmotic urine, suggesting that it has never been under selection pressure in fresh water.

In terms of overall water balance, terrestrial amphipods appear to be less well adapted to desiccation stress than the littoral forms. Nevertheless, since they are not adapted to fresh water, it must be deduced that they emerged directly from the sea on to land. This movement appears to have been accomplished, as in the isopods, more by behavioural adaptation and the retention of tolerance mechanisms than by physiological regulation.

### 4.7.3   *Decapods*

Many families of decapods contain terrestrial species, some with close marine relatives, and some with their closest relatives in fresh water. Here we will compare the superfamilies Potamoidea and Parathelphusoidea, which contain crabs associated with fresh water, with the families Ocypodidae and Grapsidae, which contain semi-terrestrial crabs mostly with a marine origin.

The 'freshwater crabs' of the superfamily Parathelphusoidea are found in India, Indonesia and Australia, while those in the superfamily Potamoidea occur in Europe, Africa, the Middle East and the Far East. *Potamon niloticus* is an aquatic species of the Potamoidea, common in lakes and rivers in eastern Africa. It differs from many other freshwater animals in that it has a low permeability of the body surfaces to salts and water, whereas other animals have a low permeability to salts but a high permeability to water. It is therefore able to survive in fresh water while maintaining haemolymph with a high osmotic pressure ($492 \text{ mOsm.kg}^{-1}$), and producing very small quantities of isosmotic urine (Shaw, 1959). The small amount of salt lost by diffusion is replaced by active uptake. This situation contrasts with that of the freshwater amphipods because of the size difference: freshwater amphipods actually have a lower permeability to water than the larger crabs, but because of their enormously larger surface:volume ratio, they all have to produce hypo-osmotic urine in order to eliminate excess water. Like the freshwater amphipods, freshwater crabs are thus osmotic regulators, but they cannot withstand much change in internal osmotic pressure. In *Potamon niloticus*, the normal osmotic pressure of the haemolymph is $492 \text{ mOsm.kg}^{-1}$, and most individuals can withstand internal values of about $550 \text{ mOsm.kg}^{-1}$; but only a few individuals can tolerate an increase to over $730 \text{ mOsm.kg}^{-1}$ (Shaw, 1959).

On land, the 'freshwater crabs' in general do not have a very high rate of evaporative water loss, but can only withstand total losses of up to 20% of their body weight (Lutz, 1969). The exception to this is the terrestrial species *Holthuisana transversa*, which appears to use the same basic physiological mechanisms for regulating the composition of body fluids as do the freshwater species, but can withstand much greater degrees of dehydration (Greenaway & MacMillen, 1978). It can tolerate a water loss of 31% of its body weight, and can also tolerate a change in osmotic pressure of its haemolymph from 500 to $980 \text{ mOsm.kg}^{-1}$. This capacity for tolerance differentiates it from most freshwater invertebrates, which generally regulate their osmotic pressure at approximately the normal level in all salinities up to that at which the environmental equals the internal osmotic pressure, and then show increasing mortality as osmotic pressure rises above the normal level. *H. transversa* therefore possesses mechanisms of *tolerance* that we might not have expected from an animal with an apparently freshwater ancestry.

Two alternative hypotheses could explain this apparent anomaly. One possibility is that freshwater ancestry has in this case led to some pre-adaptations to life on land. One of these pre-adaptations is the low permeability of the epidermis, associated, as explained above, with large size and small surface area. If this hypothesis is correct, the development of tolerance is surprising, but it could be a later acquisition allowed by the relative impermeability – much as in the higher terrestrial vertebrates. Such a development would depend upon the strong selective advantage of tolerance, in combination with physiological lability. The case of *Potamon* has shown that such lability exists, and Barnes (1968) has emphasized the variation in osmotic pressure of the blood found in a population of the estuarine crab *Australoplax tridentata*. Nevertheless, the development of tolerance must, on this hypothesis, still be seen as an exception to the rule that a freshwater ancestry engenders regulative ability, while marine ancestry leads to tolerance.

The second hypothesis assumes a more complex evolutionary history of this group of crabs. It is possible that the high osmotic pressure of the blood of *H. transversa* and the other 'freshwater' crabs represents a marine ancestry. On this hypothesis, some of the freshwater crabs, including perhaps *H. transversa*, evolved directly from marine ancestors. After a period of terrestrial or semi-terrestrial life, they secondarily invaded aquatic environments, but chose fresh waters and not the sea. In this scenario, the present-day freshwater crabs form a parallel with the freshwater pulmonate molluscs discussed in Section 4.8.1.

One of the reasons that make it hard to evaluate the relative merits of these two hypotheses is our lack of knowledge of the physiology of freshwater crabs. It is to be hoped that examination of more species, particularly perhaps the Trinidadian species *Pseudothelphusa garmani* (9.2.2) will lead to more insight on the subject.

Crabs in the families Ocypodidae and Grapsidae are primarily marine and estuarine, and most species live intertidally, but many can be termed semi-terrestrial. Grapsids such as *Cyclograpsus lavauxi* live at the very top of shingle shores, while *Aratus pisoni* is essentially arboreal in mangrove swamps. Ocypodids such as *Uca* spp. and *Ocypode* spp. are active mainly while emersed at low tide, remaining protected in burrows when the tide covers the mud flats and sand flats which are their feeding grounds. Most of these intertidal and supratidal decapods show quite different responses to osmotic stress from those shown by the freshwater crabs. Many grapsids and ocypodids, and the marine hermit crabs, can tolerate wide ranges of internal osmotic pressure. *Hemigrapsus oregonensis*, for example, can tolerate values between 600 and 1500 mOsm.kg$^{-1}$. Most, such as *H. oregonensis* and *Pachygrapsus crassipes*, the common shore crab of N. America, can hyper-osmoregulate in dilute sea water, and can perform some hypo-osmoregulation in concentrated media (Fig. 4.5). *P. crassipes*, however, also uses behavioural methods of osmoregulation: if it is allowed to visit both saline and fresh water, it can maintain its haemolymph with normal levels of sodium by regulating the lengths of time spent in each salinity (Gross, 1957).

A further group of marine and semi-terrestrial crabs have more precise physiological mechanisms for regulating the osmotic pressure of their haemolymph. *Grapsus grapsus*, for instance, can regulate its osmotic pressure within narrow limits over an external range from 50–150% sea water. *Ocypode ceratophthalma* shows a change in the osmotic pressure of the haemolymph equivalent to that of only 10% sea water when the external salinity changes from 50 to 170% sea water (Fig. 4.5). This osmoregulation is not carried out by the excretory antennal glands, as might be expected, as the urine is always isosmotic with the haemolymph. In some of the grapsids, there is evidence that the diverticula of the gut may be involved, but surprisingly, no details of the presumed extrarenal osmoregulating organs have ever been described. The antennal glands themselves are important in controlling the ionic composition of the haemolymph, if not its osmotic pressure. Although they produce isosmotic urine, they are responsible for the differential excretion of magnesium and sulphate. A good example is provided by *H. oregonensis*, in which haemolymph magnesium concentration is only 23.5 mM.1$^{-1}$, while urine magnesium concentration is 130.5 mM.1$^{-1}$. Similarly, in *P. crassipes*, haemolymph magnesium is only 10 mM.1$^{-1}$, while urine magnesium is 118 mM.1$^{-1}$.

Perhaps the most striking part of the ability to osmoregulate by semi-terrestrial crabs is their capacity for hypo-osmoregulation. This ability is uncommon in the animal kingdom, but is widespread in the semi-terrestrial grapsids and ocypodids. Gross (1964) further pointed out that in this group there is a strong correlation between the ability to hypo-osmoregulate and terrestrialness. He suggested that the ability evolved in areas of widely fluctuating salinity, such as coastal lagoons, which would have produced

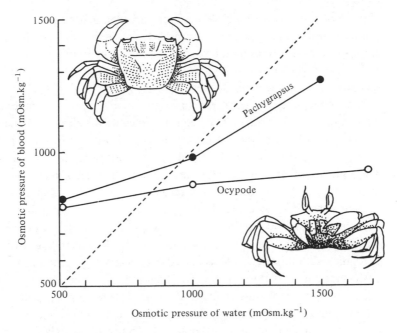

Fig. 4.5. Osmotic regulation in two intertidal crabs. The grapsid *Pachygrapsus crassipes* (filled circles) can hyper-osmoregulate, but shows only slight ability to hypo-osmoregulate. The ocypodid *Ocypode ceratophthalma* (open circles) maintains an extremely constant osmotic pressure over a wide range. After Gross (1964).

selective pressures favouring hypo-osmoregulation. It should be emphasized here that the physiology of decapod lines moving from sea to land differs markedly from that of molluscan lines in the same process. In decapods, many semi-terrestrial species have developed marked osmotic tolerance, as have the semi-terrestrial marine gastropod molluscs. The *most* terrestrial decapods, however, probably those evolving in coastal saline lagoons, have replaced tolerance with physiological regulation of osmotic pressure using hypo-osmoregulation. This the molluscs have entirely failed to do, and as a consequence they still depend upon mechanisms of tolerance. One may ask why this difference occurs. The major factor is probably *not* to do with efficiency of excretory organs, as molluscan organs appear to be quite as good as crustacean ones, at least in producing hypo-osmotic urine and therefore in effecting hyper-osmoregulation. Rather it is probably to do with the development of a relatively impermeable integument, which has prevented tremendous water loss to the outside in crustaceans. With this development, coupled with the relatively large size of decapods, it became possible for decapod organs to regulate effectively against an internal rise in osmotic pressure. In the molluscs, no production of hyper-osmotic urine could have counteracted the tremendous water loss across the body surface in hyper-osmotic conditions, and tolerance has remained the only feasible strategy.

The dependence of semi-terrestrial decapods upon an impermeable integument makes it difficult to separate lines with a direct marine origin from those with a freshwater ancestry by looking at tolerance vs regulatory ability as was done in molluscs. However, although in both cases the development of an impermeable integument has been accompanied by great osmoregulatory ability, only in *marine* forms habituated to coastal lagoons has there been the development of *hypo*-osmoregulation. This characteristic should therefore be a good guide to marine, as opposed to freshwater, origins; and it is here that the semi-terrestrial grapsids and ocypodids differ from *Holthuisana*, the terrestrial parathelphusoid with freshwater relatives discussed at the beginning of this section: although *Holthuisana* can tolerate changes in osmotic pressure, and can hyper-osmoregulate, it cannot *hypo*-osmoregulate. It should also be pointed out here that in marine littoral decapods the ability to withstand increasing osmotic pressures is better developed than the ability to withstand dilution – evaporation being more important in the littoral zone than dilution. This is reflected in the ability of most terrestrial crustaceans, such as the isopods, to withstand great increases in blood concentration, and a relative inability to withstand dilution.

### 4.8     Marine vs freshwater routes: pulmonate gastropods

Of the prosobranch gastropod molluscs discussed earlier (4.4), terrestrial lines with freshwater ancestors have retained a low osmotic pressure, and little tolerance of desiccation, while lines with direct marine ancestry have high osmotic pressure and great tolerance of desiccation. The terrestrial pulmonate gastropods also have high values of osmotic pressure – up to 400 mOsm.kg$^{-1}$ has been recorded in active slugs (Bailey, 1971). The high values achieved probably represent a marine, or at least a brackish-water origin.

#### 4.8.1     Aquatic pulmonates

The situation is complicated, at least superficially, by the presence of many fairly primitive pulmonates (the Basommatophora, those with eyes at the base of the tentacles) in fresh water, while most of the more advanced forms (the Stylommatophora, those with eyes at the tip of the tentacles) are found on land. There are, however, some Basommatophora in the marine and estuarine intertidal zones, and it is thought that these are more primitive than their freshwater relatives, which may have been secondarily derived from semi-terrestrial lines. Some of these marine Basommatophora, such as the family Siphonariidae, have colonized rocky shores, and have adopted a limpet form. This has modified them considerably from any possible ancestral form, and they will not be considered here. Instead, we shall concentrate upon two families that have retained a large number of primitive features, the Ellobiidae and the Amphibolidae.

The family Ellobiidae was proposed by Morton (1955) to contain the most primitive of present-day pulmonates. It has representatives in the rocky marine littoral, adapted for crevice life; in the supratidal regions of estuaries; in terrestrial habitats fringing tropical coastlines; and in inland terrestrial regions of the temperate zone. Of most interest in the present context are the supratidal estuarine species, and the coastal terrestrial species. The supratidal estuarine category contains most of the forms found in Europe and North America. Typical of these is *Ovatella myosotis*, found high up on saltmarshes, often under thick vegetation (see 6.1.2). It can withstand very large changes in the salt concentration of the mud on which it lives – from fresh water to $90^o/_{oo}$, or nearly three times sea water (Seelemann, 1968). Like most estuarine molluscs, it does not osmoregulate, except at low salinities, but it does maintain its blood very hyper-osmotic to all media – possibly an adaptation allowing it to absorb water osmotically from the substrate in order to maintain body turgor. It is not known how *O. myosotis* maintains an osmotic difference from external media. In other parts of the world different genera occupy equivalent positions on saltmarshes. In North America the commonest species is *Melampus bidentatus*, which when desiccated can survive a loss of nearly 80% of its body water (Price, 1980). The saltmarsh forms are thus seen to be physiologically adapted to allow them to tolerate the harsh conditions of the upper shore, and particularly the very high and very low salinities found in these regions.

The coastal terrestrial ellobiids have been neglected in terms of both ecology and physiology, although they reach quite large sizes and may be significant components of maritime ecosystems in the tropics. *Pythia* spp., for example, are common in tropical forests bordering the coastline in the Indo-Pacific, and reach shell lengths of over 2 cm. It is important that these, and the larger mangrove ellobiids such as *Ellobium aurisjudae*, which grow to shell lengths of over 7 cm, be investigated. In the meantime, the most likely hypothesis is that ellobiids have become terrestrial with little physiological adaptation except the development of extreme tolerance mechanisms: very much a parallel to the situation found in the pomatiasid prosobranchs (see 4.4.1). This adoption of tolerance must be coupled with appropriate behavioural responses which allow the snails to select suitable micro-habitats, but these may well be similar to those in the high-shore marine and estuarine forms.

The above hypothesis is reinforced by evidence from a second primitive basommatophoran family, the Amphibolidae (see also 5.1.2). This family is restricted to the Indo-Pacific, where its members are found on estuarine mud-flats, and in saltmarshes fringing the terrestrial edges of mangrove swamps. The adult snails of the New Zealand species *Amphibola crenata* can hyper-osmoregulate in low salinities, but withstand changes in salinity from approximately 25% sea water up to 100% sea water by tolerance mechanisms: the osmotic pressure of the blood rises in proportion to that of the external sea water (Little, Pilkington & Pilkington, 1984). Even in low external salinities, the hyper-osmoregulation that occurs is not carried out by the kidney, as the urine has been shown to be isosmotic with the blood. Some extrarenal process is therefore involved, forming another parallel with the terrestrial pomatiasid prosobranchs.

### 4.8.2    *Terrestrial pulmonates*

The dependence of semi-marine ellobiids and amphibolids on mechanisms of extreme tolerance to changes in salinity is very similar to the situation found in terrestrial pulmonates, the stylommatophorans. Terrestrial stylommatophoran slugs and snails can lose large proportions of their body water by evaporation and by loss of mucus (for reviews, see, e.g. Burton, 1983; Riddle, 1983). The responses to water loss in snails are different from those in slugs, and we will consider snails first. Early work (Howes & Wells, 1934) suggested that regular periodic fluctuations of as much as 50% body weight occur in young *Helix pomatia*. While regular cycles have not been confirmed, natural variations in body water content give rise to associated changes in the concentration of sodium in the blood which may vary from 46 to 129 mM.$1^{-1}$ (Burton, 1965, 1983). Extremes of tolerance are probably shown by desert snails, which can remain inactive, but alive, for many years, but very few of these species have been investigated in detail. Most is known about *Sphincterochila* spp., which are active for only short periods each year, and remain for the rest of the year in a dormant condition (e.g. Schmidt-Nielsen *et al.*, 1971; Steinberger *et al.*, 1981). These snails, however, have also developed mechanisms for reducing the rate of water loss to a very low level. For instance, it was calculated that, if *S. boissieri* could withstand a loss of 50% of its body water, it could survive for about 4 years. Such efficiency of water retention in soft-bodied animals is perhaps surprising, but similar mechanisms are known to exist in non-desert snails such as *Helix aspersa* (Machin, 1975). When this species is inactive, the mantle collar reduces the rate of evaporation to 0.17 mg/g body weight/hour, and when the animal secretes an epiphragm, the rate falls to 0.04 mg/g/h. In *S. boissieri*, the rates are approximately half those in *H. aspersa*. As yet, the physiological mechanisms involved are largely unknown.

Variation in the water content of slugs is still more extreme, since they are not protected to any extent by external shells. Slugs may lose up to 17% of their body weight in an hour, even in a saturated atmosphere (Dainton, 1954), and the recorded variation in blood composition is enormous. In *Arion ater*, osmotic pressure of the blood varies from 97–231 mOsm.kg$^{-1}$ (Roach, 1963). In *Limax maximus*, Prior *et al.* (1983) showed that a 33% loss of weight produced a rise in the osmotic pressure of the blood from 140 to 200 mOsm.kg$^{-1}$. However, slugs also have means of reducing their rate of water loss. They have no physiological mechanism like that in the mantle of snails, but by huddling together and by keeping the pneumostome closed to a greater extent, *L. maximus* can reduce the evaporative loss of water considerably (Prior *et al.*, 1983).

The tolerance mechanisms of stylommatophoran slugs and snails, developed to a degree which differentiates them markedly from terrestrial prosobranchs of freshwater origin, are supplemented by further mechanisms which allow some regulation of the composition of their body fluids. These are considered in Chapter 9. The similarities between terrestrial stylommatophorans and marine basommatophorans suggest a direct marine origin for the terrestrial forms. However, if Solem (1985) is correct in denying a basommatophoran ancestry for the land snails, the similarities cannot be expected to show a relationship in detail; they merely demonstrate the overall pattern.

### 4.8.3 Tolerance mechanisms

Emphasis has been placed upon the importance of tolerance mechanisms in the evolution of terrestrial pulmonate gastropods from marine ancestors, in the same way as for terrestrial crustaceans with marine ancestors, but nothing has so far been said about what these tolerance mechanisms might be. This is essentially because little is known about the subject, but it is appropriate to end this section with some discussion of tolerance mechanisms in the pulmonates. It is generally accepted that, except in specialized tissues, the osmotic pressure of the tissues of most animals equals that of the blood. In both marine and terrestrial pulmonates, it has been shown that the osmotic pressure of the blood can vary enormously, either by loss of water, or by addition of salts. This osmotic pressure is produced almost entirely by inorganic ions, and in aquatic species these ions have concentrations in the blood very similar to those in the external medium. It follows, then, that 'tolerance' must be a cellular phenomenon. It is therefore most fortunate that we have some information about how the cells of a primitive marine pulmonate, *Amphibola crenata*, respond to changes in external osmotic pressure (Shumway & Freeman, 1984).

First, it has been shown that cells in the foot of *A. crenata* can perform some degree of volume regulation. Changes in cell volume would be expected if the osmotic pressure of the blood changed. For instance, if the osmotic pressure of the blood decreased, water would move into the cells, increasing cell volume, and decreasing cellular osmotic pressure. In practice, the volume of the cells in *A. crenata* does not change as much as predicted. This regulation of cell volume is widespread in euryhaline invertebrates but is not found in stenohaline forms (Lange, 1972; Little, 1981). Volume regulation probably, therefore, contributes towards tolerance.

Secondly, it has been shown that as external osmotic pressure falls, the concentrations of intracellular amino acids in *A. crenata* fall proportionately; in 100% sea water, muscle tissues of the foot contained 515 $\mu Mg^{-1}$ dry weight, whereas in fresh water they contained only 160 $\mu Mg^{-1}$. This change in amino acids accounts for a large fraction of the change in intracellular osmotic pressure, and is known as isosmotic intracellular regulation, because it allows the cells to remain isosmotic with the blood. Such regulation is well known in a variety of euryhaline invertebrates (e.g. Pierce, 1982, Lange, 1972), and is probably the basis for the volume regulation described above. Cell volume depends essentially upon the ratio of intracellular: extracellular osmotic pressure, and, if these are maintained at equal values, the volume will remain constant. The use of amino acids, and other small organic molecules, to regulate intracellular osmotic pressure allows the intracellular concentrations of inorganic ions to remain low – an essential requirement for the normal function of most cells. Isosmotic intracellular regulation therefore probably provides another component of the phenomenon known as tolerance.

The situation is not as clear as the above suggestions might imply, because we have very little information about the use of amino acids in isosmotic intracellular regulation by terrestrial pulmonates. The only detailed investigation is that by Wieser & Schuster (1975) on free amino acids in *Helix pomatia*. They found that, although some intracellular amino acids increased in concentration when the osmotic pressure of the blood increased, other organic components (ninhydrin-positive substances) decreased, approximately compensating for the change. Some anomalies remain to be explained, however, such as the very low arginine levels reported by Wieser & Schuster: these low levels contrast with the demonstration of very high levels of arginine in *Helix aspersa* (Campbell & Speeg, 1968). Furthermore, total changes in intracellular osmotic pressure were greater than could be accounted for by changes in the measured amino acids. Evidently, the whole phenomenon of tolerance in terrestrial pulmonates, and its biochemical basis, requires further investigation. It is the thesis of this chapter that such tolerance mechanisms derive from the tolerance mechanisms of marine ancestors, but it must be admitted that this derivation has not yet been proved.

## 4.9    Interstitial routes: nemertines, oligochaetes and insects

Most of the animals discussed in the above section live on the surfaces of substrates. Although some of them live in burrows, or spend much of their time hidden by organic detritus, most are exposed directly, to some degree, to the overlying air or water. Animals that live truly within the sediment, in the interstitial environment, often face quite different physical characteristics. Because sediments act as buffers to changes in salinity of overlying water, for instance, the gradations from marine to freshwater soils and marine to terrestrial soils are more gradual than the changes experienced above the surface of the substrate. In these interstitial transition zones, the selection pressures for tolerance will probably not be as great, therefore, as on the surface; but in the transitions to both fresh water and to land the final osmotic adaptation must be to dilution. It follows, therefore, that animals that are essentially adapted to the soil are likely to have evolved good mechanisms for regulating their salt and water content, in parallel with those shown by freshwater animals. If this line of argument is correct, it will be apparent that it may be difficult to distinguish between the mechanisms of osmoregulation and water balance in those soil animals invading land via fresh water, and those invading via marine intertidal sands.

Nevertheless, it is interesting to review the ways in which salt and water balance is achieved in interstitial faunas. In this section we consider the nemertines and the oligochaetes, and also the insects, since it is argued here that the insects may have had an interstitial origin.

### 4.9.1     The nemertines

The majority of nemertines are marine, but some species, such as the common *Lineus ruber*, can live in brackish water, and some genera, e.g. *Potamonemertes* and *Prostoma*, inhabit fresh water. Many of the marine littoral forms, e.g. species of *Prosorhochmus*, live at high tidal levels, while a few of the terrestrial species live in humid habitats near the top of the shore. This distribution has suggested that terrestrial forms evolved directly from marine ancestors (Pantin, 1969), although this evolution has probably occurred in parallel several times. As will be seen, however, not all the evidence agrees with this suggestion.

In the nemertines the osmotic pressure of rhynchocoelic fluid has been measured in very few species (Little, 1983). In the terrestrial species *Argonemertes dendyi* it is 144 mOsm.kg$^{-1}$, almost identical to the value found in a freshwater species *Prostoma jenningsi* (139 mOsm.kg$^{-1}$). This certainly does not support the idea of a direct marine origin for *A. dendyi*. However, *A. dendyi* may not be typical of terrestrial species, and almost certainly some other terrestrial species such as *Pantinonemertes agricola* differ from it: while *A. dendyi* can withstand immersion in fresh water but not in sea water, the reverse holds for *P. agricola*. Such differences are probably due to the parallel evolution referred to above: from studies of anatomy and distribution of freshwater and terrestrial nemertines, Moore & Gibson (1981, 1985) have established that there are at least three evolutionary lines within the Hoplonemertini (Fig. 4.6). One of these invaded brackish waters, and produced the genus *Prostoma* in fresh water. A second invaded the marine supralittoral and then the land, producing genera such as *Argonemertes* and in this line further freshwater genera evolved from terrestrial ancestors. A third line (*Pantinonemertes*) has representatives in the littoral zone, within mangrove swamps, and on land, and also probably moved directly on to land from the sea. It is therefore apparent that the freshwater species *Prostoma jenningsi* and the terrestrial species *Argonemertes dendyi* were derived from different stocks, so that direct comparison of such factors as the osmotic pressure of their body fluids is unlikely to be particularly revealing.

To some extent at least, salt and water balance in nemertines is controlled by the action of the flame cells (protonephridia). The distribution of these organs in terrestrial and freshwater forms, and studies on their mechanism of action (Wilson & Webster, 1974; Kummel, 1975) suggest that they are

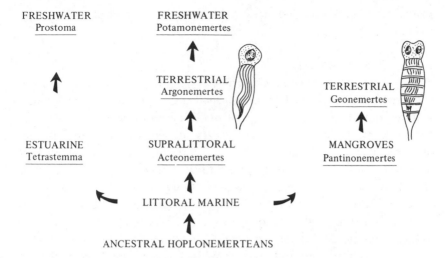

FRESHWATER
Prostoma

FRESHWATER
Potamonemertes

⬆

TERRESTRIAL
Argonemertes

⬆

TERRESTRIAL
Geonemertes

⬆

⬆

ESTUARINE
Tetrastemma

SUPRALITTORAL
Acteonemertes

MANGROVES
Pantinonemertes

⬆

LITTORAL MARINE

⬆

ANCESTRAL HOPLONEMERTEANS

Fig. 4.6. Possible sequences of evolution of terrestrial and freshwater
nemertines. The supralittoral and mangrove lines are distinguished by a
number of characters, but in particular by the structure of the excretory flame
cells. In *Argonemertes* they have a single nucleus and no strengthening of the
wall. In *Geonemertes* they consist of two cells, and also have strengthening
bars. After Moore & Gibson (1973; 1985).

concerned with the elimination of water passing in through the body
surface. The examples in which protonephridial systems have been
examined are those where they have been much expanded in comparison
with marine forms, and this expansion appears to be an adaptation to life in
dilute environments. It is difficult to imagine the selection pressure that
would produce such adaptations in surface-dwelling marine littoral
nemertines, since few other marine littoral invertebrates can osmoregulate
well, and the need to *expel* water in such habitats is usually minimal.
However, within the interstitial environment, especially in situations high
on the shore, dilution by groundwater may indeed be a problem. This is
exemplified by the situation with a New Zealand supralittoral species,
*Acteonemertes bathamae* (Moore, 1973). This species lives hidden in
crevices between damp pebbles and stones at the top of the shore, and these
are subject to drainage of fresh water from the land. *A. bathamae* has a very
well developed flame cell system, which probably is used to expel excess
water. This suggests that the marine precursors of the terrestrial forms were
well supplied with a mechanism for removing water, even without any long
period of evolution in true fresh water: frequent exposure to ground water

provided sufficient selection pressure for the enhancement of the flame cell system.

### 4.9.2    The oligochaetes

The oligochaete annelids are today found predominantly in fresh water and on land. Smaller forms, such as the aquatic tubificids and naidids, and the terrestrial enchytraeids, live essentially in the interstitial environment, while larger forms such as the earthworms form distinct burrows. The lack of large numbers of marine species may suggest that terrestrial lines have evolved from freshwater ancestors, and Brinkhurst (1982) has derived all present-day orders of oligochaetes from a haplotaxoid stem form. A freshwater origin is also suggested by the mechanisms controlling water balance in earthworms. *Lumbricus terrestris*, for example, has mechanisms very much like those in freshwater invertebrates: it can maintain the osmotic pressure of its coelomic fluid above that of dilute media, but, as the external osmotic pressure rises, it becomes an osmoconformer (Ramsay, 1949); and the hyper-osmotic regulation in earthworms is achieved partly by active uptake of chloride and partly by the production of hypo-osmotic urine (Dietz, 1974).

However, *L. terrestris* and other earthworms also have properties not found in typical freshwater animals. First, *L. terrestris* can withstand enormous changes in internal osmotic pressure: from 154 to 371 $mOsm.kg^{-1}$ (Oglesby, 1978), a range larger than would be expected for a soft-bodied freshwater invertebrate. As argued above, osmotic tolerance is a property usually incurred by animals of direct marine origin. Other earthworms are even more tolerant of high salt concentrations: *Pontodrilus matsushimensis*, which lives in sand around the high water mark on Japanese shores, can tolerate internal changes from 500 to 1200 $mOsm.kg^{-1}$ (Takeuchi, 1980b). Secondly, *L. terrestris* can reduce its integumental permeability to water when kept in air of less than 100% relative humidity (Carley, 1978), and this is likely to be a specific adaptation to the interstitial habitat. Thirdly, earthworms contain in their blood significant concentrations of amino acids – up to $13.8 \text{ mM}.1^{-1}$ in the tropical earthworm *Lampito mauritii* (Pampapathi Rao, 1963). Although we have considered the importance of amino acids inside the cells of many marine and brackish-water animals, where they act as isosmotic intracellular regulators (Section 4.8.3), the occurrence of free amino acids in blood, haemolymph and coelomic fluid is restricted in distribution. Amino acids are found in the blood of a number of marine polychaetes (Oglesby, 1978), and in the blood of other terrestrial invertebrates such as the insects, myriapods and onychophorans, but not, as far as is known, in the blood of freshwater invertebrates. No explanation is yet available for the role of amino acids in the majority of these animals, but there is one study pointing to an osmoregulatory function in a terrestrial oligochaete, *Eisenia foetida* (Takeuchi, 1980a). In this species, which withstands relatively high external salinities, rises in internal osmotic pressure are accounted for by a rise in free amino acids: as osmotic pressure rises from 150 to 250 mOsm.kg$^{-1}$, so the total amino acids in the coelomic fluid rise from 25 to 125 mM.l$^{-1}$. It is possible that other terrestrial oligochaetes utilize free amino acids in a similar way, and that the ability to form and retain extracellular amino acids has been retained from marine annelid ancestors. Lastly, the body fluids of terrestrial oligochaetes have retained a fairly high osmotic pressure. An average value for hydrated *L. terrestris* is 165 mOsm.kg$^{-1}$ (Ramsay, 1949), which compares with values of less than 100 mOsm.kg$^{-1}$ for terrestrial molluscs of freshwater origin (Table 4.3). Unfortunately, however, there appear to be no available figures for the osmotic pressure of the body fluids of truly freshwater oligochaetes with which to compare the terrestrial forms (see, e.g. Oglesby, 1978). The only investigation is that of Little (1984) on a brackish-water population of *Nais elinguis*. At an environmental osmotic pressure of 15 mOsm.kg$^{-1}$, the coelomic fluid had an osmotic pressure of 160 mOsm.kg$^{-1}$, a value very similar to that shown by *L. terrestris*. *N. elinguis*, however, showed a surprising ability to hypo-osmoregulate in salinities above 25% sea water, and this may indicate that it has specific secondary brackish-water adaptations.

In conclusion, it must be said that no clear picture of the area of origin of the terrestrial oligochaetes has yet emerged. It is essential to measure the osmotic pressure of truly freshwater species. If it should be shown that present-day terrestrial oligochaetes have significantly higher osmotic pressures than freshwater forms, the possibility of a direct origin of some terrestrial oligochaetes from marine forms might be raised. On this basis, some present-day oligochaetes might be primitively marine, and it should be noted that some tubificids appear to be fully adapted to entirely marine, even deep water habitats (Erséus, 1980; Timm, 1980). It is certainly worth bearing in mind the possibility that modern oligochaetes represent the independent invasion of terrestrial and freshwater soils from marine sediments. This point is taken further in Section 5.2.6.

### 4.9.3 *The insects, onychophorans and myriapods*

This section is in the main concerned with the Insecta (Hexapoda), but it is also appropriate to consider here the Onychophora and the

Myriapoda, as they are thought by many to have had a common origin with the insects (Manton, 1979). It is also appropriate to make some comparisons between the Insect-Onychophora-Myriapod line and the other major terrestrial taxa, the Crustacea and the Oligochaeta.

The insects, onychophorans and myriapods are terrestrial groups containing no primitively aquatic species. Insects probably evolved from myriapods, or myriapod-like ancestors, and share with them a number of characteristics, mostly related to life on land. The lines that invaded land were therefore probably myriapods and not insects – the insects evolved from them once they had become at least semi-terrestrial. It would therefore be most appropriate here to consider the osmotic characteristics of myriapods, but since so much more is known about the insects, this is where discussion will be concentrated. It is hoped that in the future more information will be available about the myriapods.

The ranges of osmotic pressure of the haemolymph found within these groups are: Insecta, 290–510; Onychophora, 180–200; Myriapoda, 150–200; Chelicerata, 360–480 mOsm.kg$^{-1}$. Most of these figures, with the exception of those for the Onychophora, are similar to those found in the Crustacea (see Section 4.7), and are considerably higher than those in the Mollusca, Annelida, Nemertea or, as we shall see in the next section, the Amphibia. While it is tempting to assign the cause of these relatively high values to a marine origin, it must be remembered, as discussed in Section 4.7.3, that decapod crustaceans which may have a freshwater origin retain high values of osmotic pressure. On the other hand, it has also been suggested that these crabs may have come direct from the sea. In any case, they are relatively large, whereas most insects and myriapods are small, and if such forms had evolved in fresh water it seems likely that their osmotic pressure would have dropped radically, as in the smaller crustaceans. The fact that it has, on the contrary, remained high strongly suggests a marine origin.

A second point reinforces the suggestion of a direct marine origin for the Insecta-Onychophora-Myriapoda line. This is the degree of tolerance to dehydration and to changes in the composition of the body fluids. These have been best studied in the insects. Although most insects are thought of as good regulators, many are also extremely tolerant of changes in the osmotic pressure of the blood. Among the Apterygota, some of the Collembola can tolerate a range from about 360 to 1360 mOsm.kg$^{-1}$ (Weigmann, 1973). Admittedly, these forms are specialized for life in saltmarshes, and little is known of the tolerance of freshwater and terrestrial soil apterygotes. However, the tolerance of some of the

Pterygota is also striking. Although in some species, such as the larvae of *Locusta migratoria*, osmotic pressure varies only between about 338 and 378 mOsm.kg$^{-1}$, in relation to feeding (Berneys & Chapman, 1974), other species can tolerate considerable change. In *Trichostetha fascicularis*, for example, desiccation caused a rise in osmotic pressure from 437 to 742 mOsm.kg$^{-1}$ (Fielding & Nicholson, 1980). The housefly *Musca domestica* showed changes from 163 to 762 mOsm.kg$^{-1}$ after feeding (Bolwig, 1953). Lethal changes have not been estimated, but many species could probably withstand greater changes than have actually been demonstrated. This is underlined by the ability of insect tissues to continue functioning in very wide ranges of osmotic pressure. Good examples are given by the Malpighian tubules of *Calliphora erythrocephala*, which secrete fluid in media from 160 to 1125 mOsm.kg$^{-1}$ (Berridge, 1968).

These extremes of ability to tolerate changes in osmotic pressure of the blood are so different from those typical of freshwater invertebrates, that their direct origin from freshwater forms seems unlikely. It may, of course, be argued that the tolerance has been developed in response to the conditions of the terrestrial environment, but this again seems unlikely because many insects have also developed systems which in practice maintain the osmotic pressure of the blood at a fairly constant level. The development of tolerance in the insects and myriapods is likely to have taken place at a time when the blood was subjected to large variations in composition. A marine littoral ancestry, rather than a freshwater one, would have provided these conditions.

The development of an excellent regulatory system, involving Malpighian tubules and rectal re-absorption, as well as the development of great tolerance, may reflect a terrestrial adaptation; but here, in principle though not in detail, the insects show a close parallel with the oligochaetes. For these, it may be argued that the interstitial or burrowing habit has brought about first the capacity for osmotic tolerance – in situations where osmotic pressure was governed by the influence of the sea and evaporation – and secondly the capacity for production of hypo-osmotic urine, the selection pressure coming from contact with fresh water in soil crevices. For the insects and myriapods, an interstitial origin may also be envisaged because of the significance of the cryptozoic niche in their life styles, particularly in those of the myriapods and the apterygote insects. To some extent, the terrestrial leaf-litter and soil habitats grade into the interstitial habitats of seas and fresh waters, as emphasized by Ghilarov (1959), providing a route for invasion by ancestral forms of the group. Some of the evidence concerning salt and water balance supports this idea. The insects show, as do the oligochaetes, the development of osmotic tolerance *and* the capacity to produce hypo-osmotic urine. Although it is usually the capacity to produce *hyper*-osmotic urine which is highlighted as a terrestrial adapt- ation in insects, some forms at least can re-absorb ions from the rectum without concomitant re-absorption of water, so that hypo-osmotic urine is excreted. Not all insects can do this – *Schistocerca gregaria*, for example, produces a rectal fluid which is slightly hyper-osmotic to the blood even when fed on a water diet (Phillips, 1964). *Calliphora erythrocephala*, however, can produce rectal fluid as dilute as 62 mOsm.l$^{-1}$ and *Dysdercus fasciatus* can produce urine with an osmotic pressure of 94 mOsm.l$^{-1}$ (Berridge, 1965), and the ability may be more widespread than currently realized.

The final parallel with the oligochaetes is the presence of amino acids in the blood, already mentioned in Section 4.9.2. These are present even in the apterygotes, and presumably form an ancient characteristic of the group. Since their concentration in the blood does not change during desiccation, however, at least in *Thermobia domestica* (Okasha, 1973), their present function is presumably not to do with osmoregulation. The parallel with oligochaetes can probably not be taken any further at present, and in any case it must be emphasized that, while the oligochaetes have remained bound to the soil environment, many insects have become emancipated from the cryptozoic habitat.

## 4.10    The freshwater route: vertebrates

From the discussion in previous sections, it may be concluded that the evidence from the distribution of animal groups in particular habitats, and the evidence from a consideration of present-day mechanisms of water balance both suggest that most terrestrial invertebrate groups have originated directly from the sea. Some have remained surface dwellers, and some have used the interstitial route, but the few groups that are thought to have come from fresh water, such as some of the prosobranch gastropods and some of the decapods, have in general not become very widespread in terrestrial ecosystems. In this they contrast completely with the vertebrates, which are thought to have used a freshwater route to land. It is therefore important at this stage to consider the evidence about the origins of the terrestrial vertebrates (see also Section 8.3).

The earliest vertebrates preserved as fossils come from the lower and middle Ordovician periods, and lived in marine environments (Halstead, 1973; Bray, 1985). During the Silurian and Devonian many forms were found in fresh water, and the early groups radiated into marine, brackish-

Table 4.6 *Blood composition of selected fishes*

| Species | Osmotic pressure (mOsm.kg$^{-1}$) | Na$^+$ | K$^+$ | Ca$^{2+}$ | Mg$^{2+}$ | Cl$^-$ | Urea |
|---|---|---|---|---|---|---|---|
| | | | | (mM.l$^{-1}$) | | | |
| Class Agnatha (cyclostomes) | | | | | | | |
| Marine | | | | | | | |
|   *Myxine glutinosa* | 962 | 558 | 9.6 | 3.2 | 9.7 | 576 | — |
| Fresh water | | | | | | | |
|   *Lampetra fluviatilis* | 253 | 120 | 3.2 | 1.0 | 1.1 | 96 | — |
| Class Chondrichthyes (elasmobranchs) | | | | | | | |
| Marine | | | | | | | |
|   *Mustelus canis* | 1011 | 288 | 8 | 5 | 3 | 270 | 342 |
| Fresh water | | | | | | | |
|   *Pristis perotteti* | — | 217 | 6.5 | 4.2 | 0.9 | 193 | — |
|   *Pristis microdon* | 548 | — | — | — | — | 170 | 130 |
| Class Osteichthyes (teleosts) | | | | | | | |
| Marine | | | | | | | |
|   *Lophius piscatorius* | 452 | 180 | 5.1 | 2.8 | 2.5 | 196 | 0.3 |
| Fresh water | | | | | | | |
|   *Salmo salar* | 328 | 117 | 2.2 | 2.4 | — | 130 | — |
| Class Osteichthyes (coelacanths) | | | | | | | |
| Marine | | | | | | | |
|   *Latimeria chalumnae* | 1181 | 181 | 51.3 | 3.5 | 14.4 | 199 | 355 |

*Source:* Little, 1983: data from various authors.

water and freshwater environments. It is therefore accepted by many that, after a marine origin, the vertebrates moved into fresh water and underwent rapid radiation. The palaeontological evidence for a marine origin of vertebrates is not, however, utterly conclusive, since it deals with armoured forms only, and the possibility remains that soft-bodied vertebrates evolved in fresh water from soft-bodied ancestors which moved in from the sea. Such a sequence was envisaged by Berrill (1955), who argued the case for the neotenous origin of vertebrates from the larvae of tunicates which exploited river detritus brought into the oceans, and then entered estuaries and rivers. Griffith (1987) suggested that the earliest vertebrates were anadromous – the bony adults living mainly in the sea, but migrating to breed in fresh water, where the soft-bodied young developed. To this debatable picture we may add the physiological evidence.

The high osmotic pressure of the blood of marine hagfish, and the fact that this is made up entirely of inorganic ions (Table 4.6), is supporting evidence for a marine origin of cyclostomes. On this basis it is reasonable to accept the lampreys as modified subsequently for freshwater life by a lowering of the osmotic pressure of the blood: even the marine lampreys have an osmotic pressure of only about 25% sea water. Griffith (1987), however, has drawn attention to two other characteristics of salt and water balance in hagfishes that could equally support a freshwater origin: their possession of glomerular kidneys (able to remove large amounts of fluid from the body) and their use of branchial ion pumps to maintain sodium and chloride balance. In his view, hagfishes could therefore be only secondarily marine.

It is interesting, then, to consider the blood composition of the other major groups of marine fish, the teleosts and elasmobranchs (Table 4.6). Teleost blood has a salt content one third to one half that of sea water, and there seems no obvious explanation of this, except that the teleosts evolved in, or at some early time became adapted to, fresh water. During subsequent recolonization of the sea they retained this low osmotic pressure, although no convincing reason why they should have done so has been proffered. However, there must have been considerable selection pressure to ensure that the blood did not become once more isosmotic with sea water. The fact that teleosts retain low osmotic pressures incidentally argues against the idea that hagfishes are secondarily marine – if hagfishes have returned secondarily to the sea, they would be expected to retain hypo-osmotic blood in the same way as teleosts. Thomson (1980) has queried the whole idea that low osmotic pressure would have been likely to evolve only in fresh water, but the selection pressure that might produce the same effect in the sea is totally unknown, and for the present a freshwater environment seems the most likely explanation.

The case with the elasmobranchs is rather more complicated. Elasmobranchs at the present time form an essentially marine group, and their fossil history suggests that they have never been involved in any freshwater stage. Yet their blood is unlike that of all other primitive marine animals except the coelacanth: it has a salt content one third to one half that of sea water, but an osmotic pressure just slightly greater than that of sea water. The difference is made up by organic molecules such as urea and trimethylamine oxide. The origin of this strange composition might well have been explained by a freshwater origin followed by recolonization of the sea, if other evidence supported it. The evolutionary sequence could then be envisaged as following the ontogenetic sequence found in the brackish-water crab-eating frog *Rana cancrivora* (Fig. 4.7), where the tadpoles have blood with a low osmotic pressure, made up of inorganic ions, but as they grow the osmotic pressure rises due to accumulation of urea (Gordon &

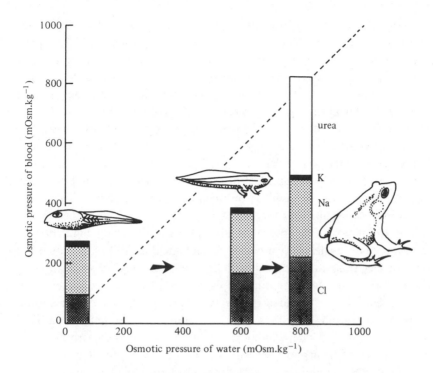

Fig. 4.7. Generalized sequence of changes in blood composition of the crab-eating frog *Rana cancrivora* during development. Eggs are usually laid in fresh water, and young tadpoles are hyper-osmotic regulators. Later tadpoles are good hyper- and hypo-osmotic regulators, but do not metamorphose in concentrations over $200 \text{ mOsm.kg}^{-1}$. Adult frogs are very tolerant of salinity changes, and are isosmotic in high concentrations: osmotic pressure of the blood is regulated to a large extent by retention of urea. The dotted line is the line of no regulation. After Gordon & Tucker (1965) and Gordon, Schmidt-Nielsen & Kelly (1961).

Tucker, 1965). This argument cannot be carried further at present, but it may perhaps serve to point out that we know very little about the habitat in which the early vertebrates evolved because we have records only of the armoured forms. It is perfectly possible that these records give us a biased viewpoint.

Freshwater amphibians have body fluids with an osmotic pressure of about $180-270 \text{ mOsm.kg}^{-1}$ (Alvarado, 1979). In general, survival is poor if this osmotic pressure rises, either from desiccation or from increasing salinity of the external medium. Amphibians are similar to freshwater invertebrates in this respect, and the similarity extends to their renal systems, which produce hypo-osmotic urine, but are unable to produce hyper-osmotic urine. There are very few exceptions to this typical 'freshwater state' in any amphibians, aquatic or terrestrial. One of the most striking is *Rana cancrivora*, mentioned above. Another is *Scaphiopus couchii*, a desert toad which can tolerate changes in blood concentration from 294 to 606 mOsm.kg$^{-1}$ during desiccation (McClanahan, 1967). As in *R. cancrivora*, most of this increase in osmotic pressure is due to accumulated urea, and this suggests that high concentrations of organic molecules may be more tolerable than high concentrations of inorganic ions. The situation is reminiscent of that in brackish-water invertebrates, where amino acids are used as intracellular osmoregulators to prevent build-up of inorganic ions (Section 4.8.3).

The ability of a few species of amphibians to tolerate high internal osmotic pressures demonstrates the physiological flexibility of the group; but the fact that the great majority can tolerate only small changes in osmotic pressure of the blood, and especially of its electrolyte content, suggests comparison of the terrestrial species with terrestrial invertebrates of freshwater origin. Like these, most amphibians are limited to areas near water, or to humid micro-habitats. However, while there are very few invertebrates of freshwater descent that have moved out of their humid micro-habitats, amphibians or their ancestors at some time gave rise to reptilian stock which has colonized the driest of micro-habitats. It is important to emphasize the properties which have allowed vertebrates to do this while the invertebrates have been confined to humid niches. The main factors here seem to be a combination of larger size and the development of a great reduction in the permeability of the skin to water, and these have allowed the reptiles to become truly terrestrial.

Neither the amphibians nor the reptiles have the ability to produce hyper-osmotic urine, from which it may be concluded that this capacity arose after the vertebrates became independent of water. This contrasts with the situation in insects, where an impermeable epidermis and hyper-osmotic urine were probably early characteristics of the insect-myriapod line. Once again, size is important here, because the amphibians were large enough to be able to tolerate long periods away from water, even without a very impermeable skin, whereas the insects, being smaller, must have developed mechanisms for both tolerance and regulation in order to be able to spend even short periods away from their cryptozoic habitat. According to Maddrell (1981), insects evolved in 'osmotically and ionically stressful environments', and this idea agrees well with the hypothesis put forward here of the invasion of land by insects (or their ancestors) via the interstitial environment, where the initial evaporation stress led to development of osmotic regulation. The apparent parallels with vertebrates in terms of the ability to live in dry environments therefore disguise basic differences in the origin and function of the physiological systems involved. The vertebrates are basically regulators, with little tolerance of internal change, having evolved in fresh water. Their development of an impermeable skin, together with their larger size, has, as it were, allowed them to move into desiccating environments *in spite of* their freshwater origin.

## 4.11    Conclusions

The main theme of this chapter has been a simple one: that terrestrial lines with a direct origin from marine ancestors have retained mechanisms of osmotic tolerance, while those with an indirect origin, i.e. via fresh water, have retained mechanisms of osmotic regulation. It is further argued that in most cases the strategy of osmotic tolerance has been more successful on land than that of osmotic regulation. This is essentially because environmental changes on land are enormous compared with those in water, and regulatory mechanisms are unlikely to be effective unless the organisms have *also* developed an efficient waterproofing layer. This development has been rare. The success of marine-derived tolerance on land is reflected in the number of terrestrial groups with a direct marine origin. These are argued to be littorinacean prosobranchs, chelicerates, isopod and amphipod crustaceans, most decapod crustaceans, pulmonate gastropods, and via the interstitial route the nemertines and possibly some oligochaetes and the insects or their ancestors. In contrast, the terrestrial groups with freshwater ancestry and associated regulatory ability are limited to the architaenioglossan prosobranchs, a few decapod crustaceans and the vertebrates.

The conclusions summarized above are, of course, tentative, but they appear to provide a coherent theme, and they also provide a means of formulating hypotheses about animal origins that can be tested by examining the physiological capacities and mechanisms of present-day species. Nevertheless, it must be admitted that conclusions reached by such methods are still only of a very general nature. To say that a particular line evolved directly from sea to land tells us little about the intermediate conditions through which the ancestral types passed, and nothing at all about how adaptations other than those for salt and water balance evolved. In the next four chapters, consideration is given to rather more specific routes on to land, and to the importance of various types of intermediate habitat for prospective terrestrial colonizers.

# 5

## Muddy and sandy shores: interstitial fauna and the burrowers

The beach was alive with hoppers feeding on the refuse, but the coarse sand was not productive of other animal life.
J. Steinbeck (1958) *The Log from the Sea of Cortez*, London: Heinemann.


In the intertidal zone of sheltered marine and estuarine regions, where wave energy is low, the dominant habitats consist of large expanses of gently sloping mudflats. These normally reach a level of about mean high water of neap tides. Above this, to approximately mean high water of spring tides, the flats are usually colonized by vegetation. Where the shore does not have protection from wave action, the deposits tend to be coarse, and the shore consists of sands, shingle or rock. This chapter will look at mud and sandflats, and the adaptations of selected animals in them, and will discuss their possible importance as routes on to land. The vegetated areas – saltmarsh and mangrove – will be covered in Chapter 6, and shingle and rocky shores will be discussed in Chapter 7.

### 5.1     Mudflats

#### 5.1.1     *Physical and biological influences in mudflats*
Many of the physical conditions in mudflats are determined by the nature of the particles making up the sediment: the proportion of various sizes of particle, particle shape, the degree of sorting and compaction, and to some extent the mineral composition of particles (Levinton, 1982; Gray, 1981). These determine how much water flows between the particles, and therefore how much oxygen is supplied to the interstitial spaces. In fine deposits, water is held in position between the mud particles, and flow rates are small so that the water is rarely replenished. This leads to deoxygenation. Where particles are coarse, water tends to move rapidly, so that oxygen supplies are adequate, but the sediment tends to dry out at low tide. This type of regime is characteristic of sand flats (5.2).

86

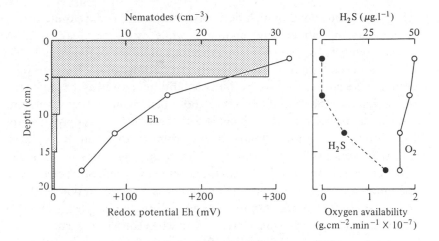

Fig. 5.1. Vertical gradients in a fine sand beach at Port Erin, Isle of Man, during summer. Median particle diameter was 179 μm. On this sheltered beach, redox potential showed a marked discontinuity, oxygen availability declined with depth, and below 10 cm, $H_2S$ was present. Nematodes were found in large numbers only in the surface layers. After McLachlan (1978).

The low rate of oxygen renewal in fine sediments (muds and fine sands) leaves only the surface layer with available oxygen. Below this surface layer, oxygen is rapidly used up by bacteria, and sulphate is reduced to hydrogen sulphide (Fig. 5.1). This has a very important consequence for animals living in mud; they live either on or very near the surface, or they construct burrows in which they maintain high levels of oxygen, independent of the surrounding sediment. The burrows in turn are of crucial importance in altering the physical properties of the mudflats, because they allow water to pass through the sediments much faster than would be possible from diffusion (Harrison & Phizacklea, 1987).

Many types of burrowing organism are sedentary, and do not concern us, since it is only the mobile forms that have moved from aquatic sediments on to land. Others emerge at low tide to feed on the surface detritus – many crabs, for instance, come up out of their burrows and sift through the fine deposits left by the receding tide. They are often accompanied for a time by deposit-feeding gastropods which move over the mudflats while they are still wet. In general, however, the molluscs tend to feed while the mudflats are covered by water, and they form temporary burrows to avoid desiccation as the flats dry out.

Intertidal mudflats provide ideal environments for such detritivores, which are often found in high densities. Marine areas have quite diverse assemblages of infauna and epifauna, while estuarine areas have low species numbers, but still contain enormous numbers of individuals. The estuarine species numbers are probably low because of the harsh physical regime, especially variations in salinity conditions, as well as the uniformity of the habitat (Barnes, 1984). Once these factors have been overcome, individual species can take advantage of the rich food supply available from production in the saltmarshes and from imports both from the sea and the rivers feeding the estuary. With these high rates of secondary production, it is not surprising that there are heavy predation pressures on the infauna of mudflats, and this predation is a vitally important factor influencing infaunal adaptation. There are three main types of mudflat predator: wading birds, which have to feed when the flats are emersed, and invertebrates and benthic-feeding fish, which feed when the flats are covered by water. All can take a heavy toll of mudflat invertebrates (McLusky, 1981; Reise, 1985). On estuarine mudflats, for instance, invertebrates such as *Crangon crangon* may remove a large percentage of post-settlement young. Wading birds take enormous numbers of the polychaete *Nereis diversicolor*, the amphipod *Corophium volutator* and the bivalve *Macoma balthica*, and it has been estimated that at some sites in England, about 90% of the adult macrofaunal biomass may be removed by waders and ducks. In other regions, fish predators may be more important than the birds, and the juveniles of many fish species migrate into estuaries to feed on the rich invertebrate fauna of the mudflats.

The combination of physical influences such as degree of desiccation and availability of oxygen, together with biological influences such as the timing of predation, set limits to the times at which the infauna can be active. In fact, periods of activity on mudflats are closely governed by the rhythm of the tides, as suggested above. The flats are covered nearly every high tide, so that the animals are subjected to a regular alternation of immersion and emersion, and are seldom left out of water for as long as 12 h. The physiological, structural and behavioural adaptations of mudflat animals are therefore geared to short-term aerial exposure, with regular replenishment of water supplies. Those that feed when the flats are emersed risk danger from desiccation and bird predation, while those that feed during immersion risk predation by fish and invertebrates but are limited to an essentially aquatic life. Here we will take a few examples from the group active under aerial conditions, to investigate the adaptations necessary for such a life from ancestors active only under water.

### 5.1.2 *Pulmonate gastropods:* **Amphibola**

On estuarine mudflats in New Zealand, one of the commonest detritivores is the primitive pulmonate gastropod *Amphibola crenata*. This species belongs to the family Amphibolidae, which is restricted to mudflats and saltmarshes in the Indo-Pacific. *A. crenata* has no gill, but a vascularized mantle cavity that acts as a lung. It can breathe either air or water, and its rate of oxygen uptake remains unchanged when it is submerged. It produces free-swimming veliger larvae (Little *et al.*, 1985), so, although it lives high up in the intertidal zone, it is inescapably linked to the sea: it forms spiral egg masses with a mixture of mud and secreted mucus, and the larvae hatch directly from these masses. *A. crenata* tolerates enormous changes in salinity without osmoregulating, but maintains its blood slightly hyper-osmotic to the external water in low salinities (4.8.1). It feeds during emersion by taking in surface detritus as it crawls across the mudflats, leaving behind it a continuous faecal string. If the surface becomes too dry, the snail burrows by pushing the leading edge of the shell under the mud. It also burrows at high tide. This activity pattern, geared to the period of emersion, may be related to factors such as predation pressure, but it is more likely that it points to air breathing being the preferred way of obtaining oxygen. Indeed, there is some suggestion that the ancestors of *Amphibola* may have lived even higher on the shore, and that it was at this time that the pulmonate stock lost the gills and acquired a lung. Both *Amphibola* and the other primitive pulmonates such as the Ellobiidae, to be discussed later (Section 6.1.2), have ciliated ridges inside the mantle cavity which circulate the water through it. In terms of ventilation, these ridges therefore effectively replace the cilia of the gills in other snails such as the prosobranchs (Pilkington, Little & Stirling, 1984).

There is no doubt that *Amphibola* is adapted for a life on very damp (saturated) sediments by the combination of anatomical, physiological and behavioural characteristics described above. It needs these damp sediments to feed and to construct its egg masses, and as the sediment dries out it burrows and becomes inactive. Nevertheless, it can breathe air as well as water, and its tolerance mechanisms allow it to be active for a large proportion of the emersion period. These abilities mark a large difference from subtidal animals, and the fact that *Amphibola* burrows and becomes inactive at high tide is another clear distinction. Emergence into environments remote from the influence of the tides is, however, prevented by the tie to saturated sediments, the requirement for a constant salt supply, and the dependence upon the production of pelagic larvae. These characteristics are common in many littoral marine animals, and mark a 'halfway house' in the process of the invasion of land. Other pulmonates, living higher up in the intertidal zone, have carried the adaptation process further (Section 6.1.2).

### 5.1.3    *Prosobranch gastropods:* Hydrobia ulvae

Although the pulmonate *Amphibola* described above is common on mudflats in New Zealand, the majority of snails found on marine and estuarine flats are prosobranchs. Many families are represented, but one of the most widespread is the Hydrobiidae, found in estuaries throughout the world. One of the best known is the European species *Hydrobia ulvae*. This is only about 8 mm long, but reaches densities of over $100,000.m^{-2}$. It is active while submerged and while the mud surface is damp, but as the substrate dries out the snails either burrow or attach themselves to plant stems with mucus and close the aperture of the shell with the operculum. Unlike the pulmonate *Amphibola, Hydrobia* has a gill and the cilia on this circulate water through the mantle cavity, thus providing oxygen for respiration as well as removing excretory waste. *H. ulvae* is not adapted for air-breathing. Its tolerance of salinity change is great, however, as befits an estuarine animal, and it can withstand both dilution and desiccation.

*H. ulvae* has been described as a detritus feeder, but most of its energy supply is probably derived from digesting diatoms rather than bacteria or dead organic material, and it is in reality a grazer. Its behaviour is to a great extent governed by the search for more surfaces to graze (Barnes, 1981), and populations on saltmarshes are seen to move on to the stems of such saltmarsh plants as *Salicornia* when these are covered by the tide (Fig. 5.2). As soon as the sediments or plant stems dry up after the recession of the tide, *H. ulvae* becomes inactive, probably in direct response to changes in water cover rather than being governed by intrinsic rhythms (Barnes, 1986). Like *Amphibola*, therefore, it is adapted to life on intertidal flats, and it requires saturated sediments for activity. Unlike *Amphibola*, it is active both in water films and when submerged. Although its adaptations differ from those of *Amphibola*, they fit the snail for limited periods of aerial exposure: in the case of *Hydrobia*, aerial exposure can be said to be 'tolerated', and activity is suspended until the tide returns.

The closest relatives of *Hydrobia* are found in fresh water, and only a few species of hydrobiid (e.g. *Geomelania*) are found on land. The route by which these terrestrial representatives emerged is not known, but, from consideration of *H. ulvae*, it seems unlikely that they would have made the direct step from marine or estuarine mudflats. Of the more than 1000

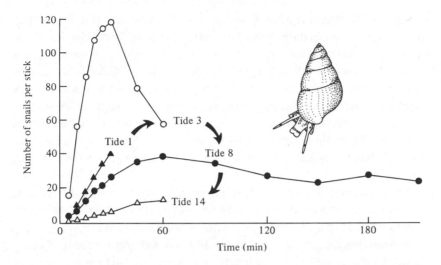

Fig. 5.2. Climbing activity of the estuarine snail *Hydrobia ulvae*. The graphs show the numbers of snails climbing up experimental sticks in the laboratory, during four tides. The tides are numbered in the sequence of occurrence after the first one to cover the snails. Climbing activity is greatest early in the sequence, because while climbing the snails are feeding, and before the tidal sequence began, they were effectively starved. After Barnes (1981).

described species, many are found in brackish waters and especially in lagoons. It is possible that coastal brackish-water lagoons formed a route to land, because here terrestrial vegetation may occur right to the water's edge. They may thus form a parallel to the situation in saltmarshes considered in Chapter 6, and indeed many lagoons are fringed by saltmarshes or mangroves.

### 5.1.4 *Crabs: the grapsid* Helice *and the ocypodid* Macrophthalmus

Various species of crab are common on mudflats and sandflats throughout the world. Two species that have been particularly well studied are *Helice crassa* (family Grapsidae) and *Macrophthalmus hirtipes* (family Ocypodidae), both common on marine and estuarine flats in New Zealand. The two provide an interesting comparison between a species adapted to low-water flats (*Macrophthalmus*) and a species adapted to the top of the shore (*Helice*).

Until recently it was thought that both species were active mainly at low water, but it has now been shown (Williams, Naylor & Chatterton, 1985) that they are also active at high tide. At this time they make wide-ranging exploratory movements away from the burrow, whereas at low tide they spend their time feeding close to their burrows. Both species show endogenous rhythmic activity patterns geared to the tides when examined in the laboratory, but these are not always so clear in *Helice*. This may be because there are often periods when high tides do not reach the levels occupied by *Helice*, so that the rhythmicity lacks reinforcement. Nevertheless, both species are normally active on the shore at low water, but at different tidal heights. The adaptations governing this vertical separation (Fig. 5.3) have been examined in a series of papers (e.g. Jones & Simons, 1982).

*Helice* normally lives in well-drained, compacted sediments above mean tide level, where it is exposed to the air for more than 6 h in each tidal cycle. It feeds by picking up dry sediment with its chelae and then passing this to the mouthparts for sorting. Consequently it is not efficient when feeding on the wet badly drained substrates of the mid-shore. It also forms permanent burrows down to the water table, which it uses as retreats from desiccation and from predators, and which it defends. These burrows are stable in high-shore sediments but would be difficult to maintain in wet sediment. In contrast, *Macrophthalmus* has two feeding mechanisms, using its chelae in dry sediments, and the setae on legs and chelae in wet sediments. While it, too, burrows into the sand and mud, the burrows are temporary and they are not defended.

Other physiological adaptations concern osmoregulation and water balance. Both crabs can perform hyper-osmotic regulation in low salinities, but are isomotic in salinities near sea water. *Helice* is somewhat better adapted to reducing water loss in air than *Macrophthalmus*, but it is unlikely that this difference is significant in the natural habitat. As far as these two adaptations are concerned, therefore, there is little difference between the two species. In respiratory terms, however, *Helice* is much better adapted to breathing in dry habitats than is *Macrophthalmus*. Both species have very similar gill areas and structures, but they differ in the way that branchial water is utilized (Hawkins & Jones, 1982). In *Helice* the water that is circulated over the gills passes out of the exhalant openings and then runs in a thin film over the *exterior* surface of the gill chamber before re-entering the chamber itself. Water is held on the surface of the body by a layer of setae. The water is thus re-oxygenated, and can be used again for respiration. It may also aid in evaporative cooling, and it can be used to allow the mouthparts to continue sorting food. In *Macrophthalmus hirtipes*, the water emerging from the exhalant openings is not retained by layers of setae, and is therefore lost. This species cannot respire and feed in

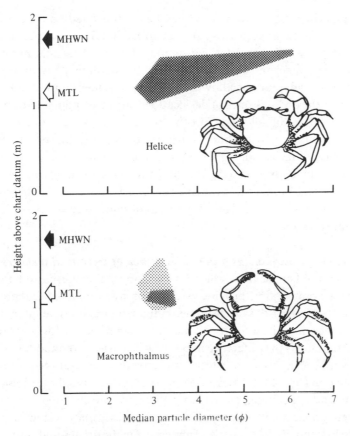

Fig. 5.3. Distribution of two New Zealand crabs in relation to particle diameter of the sediment and tidal height. Light stipple shows 1–9 crabs found per search; dark stipple shows 10 or more crabs per search. *Helice crassa* is found higher on the beach and in finer sediments (larger φ values). *Macrophthalmus hirtipes* is found lower down and in a more restricted range of coarse sediments (smaller φ values). φ = −log₂ of the particle size in mm, so for example a φ value of 3 = 0.125 mm. After Jones & Simons (1982).

air for more than about an hour. Other species of *Macrophthalmus* may, however, be able to re-circulate water externally.

In places, these two crabs are found where intertidal mudflats contain a significant amount of sand, and here the flats may continue to levels near mean high water springs without merging into saltmarsh. In this situation, *Helice* may be found living above the tidemarks. Where a suitable substrate continues above the intertidal zone, therefore, *Helice* can become semi-terrestrial, always providing that its burrows reach down to ground-water. Of the two species, it is the better adapted to a semi-terrestrial life.

This comparison between two species has emphasized the different adaptations that fit each species for a particular habitat. Each can be said to have a 'strategy' which consists of a combination of these adaptations fitting particular environmental conditions. It is salutary to observe, however, that the adaptations allowing furthest movement on to land are not in this case the obvious ones of desiccation tolerance and regulation of the composition of body fluids. Both species have a fair degree of desiccation tolerance, and both are fairly good osmoregulators; but the critical adaptations have to do with feeding and burrowing, and with water retention in conjunction with respiration. In the case of these and other crabs, the high sand content of the substrate is very important in allowing movement to high tidal levels. The adaptations of some truly sand-dwelling crabs are taken further in Section 5.2.10.

### 5.1.5    *Summary: mudflats as a potential source of terrestrial invaders*

True mudflats seldom reach to the top of the shore, and they therefore at the present time provide no direct pathway on to land: there is usually a further barrier of sand, shingle, saltmarsh or mangrove intervening between them and the terrestrial zone. A further difficulty for potential invaders of the land is the difference in biological character of mudflats and terrestrial surfaces: apart from those with growths of sea-grasses, many mudflats have an almost total lack of plant cover, and to avoid the physical regime of the macro-climate, animals must burrow into the mud. There is no equivalent on the mudflat of the cryptozoic niches found on land, which are provided essentially by dead plant material. On the other hand, some of the physical properties of mudflats are more benign than those of terrestrial soils: water supply is seldom a problem on the flats because of the regular tidal inundation, whereas water supplies on the adjacent land may be irregular.

In relation to these differences between the properties of mudflats and those of terrestrial soils, the behavioural, anatomical and physiological adaptations of present-day mudflat animals to life in air are relatively undeveloped. Although these adaptations are highly instructive in demonstrating the ways in which individual systems have been modified to cope with particular conditions, there are only a few cases in which mudflats have provided a route on to land. It is probable that saltmarshes, which lie immediately to the landward of the flats, are of greater importance, and these are considered in the next chapter. It should be noted, however, that before the origin of saltmarshes, *i.e.* before the Cretaceous, mudflats may have been more extensive, and in general reached higher up the shore. Early mudflat animals may conceivably have become better adapted to aerial life than those of present-day flats.

## 5.2    Sandy shores

The last section was concerned with coasts where wave energy is low, and the beach deposits tend to be fine silts and clays. On exposed shores, mudflats are replaced by sandflats. In temperate zones, areas of sand dunes usually lie to the landward of extensive sand beaches. These dunes are derived from the sand at high levels on the beach, which dries out at some stages of the tide, and may then be moved inland by the action of the wind. The majority of tropical shores, in contrast, are *not* backed by dune systems (Pethick, 1984), but are colonized by fast-growing vegetation (Jennings, 1965). This may be because the sand is more often kept damp by rain, and hence is in general more stable.

In spite of the predominant juxtaposition of sandflats and sand dunes on temperate shores, the two environments have seldom been considered together. The sandflats are regarded as the province of marine biologists, while the dunes are investigated by terrestrial biologists. This unfortunately means that the junction of the two is often neglected. To some extent it must be admitted that this also reflects the relatively sudden change in conditions between flat and dune. Here we can best discuss the interface by first describing conditions within the intertidal sands, and then contrasting these conditions with those of the dunes. Following this, we will examine the transition to truly terrestrial soils.

### 5.2.1    *The physical environment of marine sandflats*

Because the distribution of particle sizes on sedimentary shores relates directly to water velocity, coarse sand beaches are found where either wave action or longshore currents are high. The water movements associated both with waves (which act normal to the shore) and with currents (parallel with the shore) are involved in sorting the sediment particles, so that fine particles are transported to areas of low current velocity. At the same time, sorting may produce considerable variation in particle size on any one beach. In particular, finer particles are often found low on the shore, and coarser ones at the top, and these differences are often also associated with an increase in slope from low shore to the top of the intertidal zone. The reasons for this distribution are far from understood, but may be related to the fact that the velocity of waves breaking on the shore tends to be greater than the velocity of water draining back down the beach after the waves have broken. This asymmetry in velocity may tend to leave larger particles higher up the shore (taken there by high incoming velocity), whereas smaller particles come to equilibrium further down the beach, moved by the lower outgoing velocity (see Pethick, 1984, for discussion).

However the sorting occurs, its presence is an important factor in affecting the environment within the sand, because sands of different particle-size composition, and with various degrees of sorting, show a wide range of physical properties. In general, the better sorted sands have more space between the particles (a higher 'porosity') than do badly sorted sands. This is because the small particles in a badly sorted sand can partly fill up the gaps between the larger particles. The degree to which the gaps are filled up depends also on several other factors, including the shape of the particles, and how well the particles are packed together. If the sand is deposited rapidly, grains tend to be badly packed, leaving a large pore space, but when deposition rates are slow, there is time for each grain to gravitate to its equilibrium position, providing good packing and low porosity.

Although the porosity of sediments is important in defining the living space available for the fauna, a more important characteristic is probably the permeability – the rate at which water can flow through the sediment. The rate of flow determines, for example, the degree to which the sand will dry out at low water, the extent to which interstitial water may be replaced by overlying water of different quality (e.g. different in salinity or temperature), and the extent to which oxygen in the interstitial water, once used up, is replaced by incoming oxygenated water. Permeability is determined not by the overall interstitial space, or porosity, but by the size of the individual pores. Well-sorted sediments with large particles will have large pores, and hence a high permeability, while sediments with small particles, or a mixture of large and small particles, will have small pores and hence a low permeability. This is well illustrated by an example of sands from a Nigerian estuary (Webb, 1957). With sand of an average grain diameter of 0.25 mm, 50 cm height of water drained through a 10 cm depth of sand in 10 min 36 sec. With an average grain diameter of 0.15 mm, the equivalent drainage time was 25 h 30 min.

The degree to which water can circulate within a sediment is further affected by the percentage of the water that is held in place between the particles by surface forces. In coarse-grained sediments, such as Nigerian sand of an average grain diameter 0.9 mm, this may be negligible, but in finer sands a significant proportion of the water is held by capillarity: for sediments with an average grain size of 0.15 mm, about 14% of the water was not free for circulation (Webb, 1957). Capillary forces are therefore an important factor in determining the extent to which sands remain damp when emersed. Again, an example from Nigerian sand is instructive. When sand of average grain size 0.9 mm was placed above a water supply, water rose to a height of 4.7 cm in the sand. With sand of average grain size 0.15 mm, the water rose by capillarity 33.5 cm.

We can crudely summarize the effects of porosity and permeability by saying that fine sands retain a high water content but allow water to penetrate through the sediment only slowly, while coarse sands allow water to enter and to leave rapidly. Sands with mixtures of particle sizes have both reduced porosity and reduced permeability. The overall result of these effects is that the water table is frequently near the surface on the lower shore, but only rises to the surface high on the shore at times of high tide. At other times the water drains away in the coarse, permeable sands, and it may be 1 m below the surface for most of the time. Water available high on the shore is mainly capillary-held water, but even this may be scarce because there is very little restriction on water flow.

The major consequence of any reduction in water flow through the sediment is a lowering of oxygen concentration, because oxygen is used up by organisms faster than it can be replaced. This can be demonstrated by examining the depth profiles of oxygen availability in a variety of sands at different seasons, as has been done by McLachlan (1978) in the Isle of Man. Here the oxygen availability decreased with depth down to 20 cm below the surface (Fig. 5.1). Although this decrease does not appear striking, the change in chemical regime is overwhelming, so that whereas oxidizing processes were dominant at the surface, reducing processes were dominant 20 cm down. This change is readily measured by recording Eh, the redox potential, which reflects the balance between oxidizing and reducing processes. The break in the redox potential profile, known as the redox potential discontinuity (RPD) occurred at shallower depths in finer sediments. The RPD was also nearer to the surface in summer than in winter, presumably because of the faster rate of consumption of oxygen by organisms at higher temperatures. The distribution of interstitial meiofauna changed as the position of the RPD changed: in finer sediments, in winter, nematodes were much commoner near the surface, whereas in only slightly coarser sediments they were common down to 20 cm. In summer, the bulk of *all* populations was restricted to the surface. The physical characteristics of intertidal sediments, then, are responsible for the chemical conditions within the interstitial water, and these in turn control the distribution of the small animals. As will be seen later, the larger animals are to a great extent independent of these conditions because they form burrows which have their own micro-climate.

### 5.2.2    The physical environment of coastal sand dunes

First it is necessary to give a brief outline of how coastal sand dunes form and evolve. Detailed accounts have been provided by Ranwell (1972) and Pethick (1984). Towards the top of sandy shores, onshore winds tend to move dry sand grains inland. This is not a uniform procedure, however, because the moving sand grains encounter the collection of litter and debris left by succeeding high tides. Around these obstacles, wind velocity varies greatly, and sand is deposited mainly on their lee side, eventually forming a small hump. Such formations may not be permanent, but, if colonized by stabilizing grasses such as *Agropyron*, they may enlarge to form embryo dunes, up to 1 or 2 m high. *Agropyron* helps both to trap more sand, and to hold it in position. Colonization by further species of stabilizing grasses such as marram grass, *Ammophila* spp., may lead to further accretion of sand, and the evolution of fore dunes, which may be up to about 10 m high. The ridge of fore dunes is not stationary, but moves inland as it grows, to be replaced by more dunes on the seaward side if the sand supply is adequate. Dunes found further inland are therefore older, and, because they contain more organic matter added by the first colonizing plants, they provide more nutrient-rich environments which lead in turn to colonization by yet further species. Wind velocities inland are also lower than those at the seaward edge of the dunes, so that the inland dunes – often known as fixed dunes – are less mobile. In some regions, however, their vegetation may be eroded to form barren areas called 'blow-outs', and here the sand may be removed down to the water table to form areas called dune slacks: low-lying, damp regions with a specialized fauna and flora.

Conditions within the sands of the embryo dunes are similar to those at the top of the intertidal zone. Water supplies are minimal, and depend essentially upon rain, and the water table is far below the surface: at the surface, water contents on dunes colonized by *Ammophila* may be as low as 5% by volume, and the water table may be many metres below. Rain water is held in the sand, well above the water table, by capillarity, and is possibly added to by water condensing as dew. In the fixed dunes, with their much higher organic content, and therefore with a greater proportion of fine particles, water contents may be much higher – up to about 30% by volume – and in the dune slacks the soils may reach saturation.

Further changes occur in the soils of fixed dunes and slacks, but again these are mainly due to the accumulation of organic matter. The pH tends to decline because rainfall washes the carbonates out of the system, and, although nutrients accumulate as organic matter accumulates, some are washed from the dunes into the slacks. Details of these changes in one set of dunes over a period of several hundred years have been discussed by Wilson (1960). It is thus the slacks that have soils most resembling those of true terrestrial situations, and the richest plant and animal communities.

### 5.2.3    The relationships of dune sands to terrestrial soils

For marine animal lines that might move on to land through sandy shores and sand dunes, it is important to consider the final step in the transition, between dune soils and inland terrestrial soils. The variety of terrestrial soils is immense (see, e.g. FitzPatrick, 1980), but it is possible to make some generalizations about the differences between most terrestrial soils and coastal dune sands. The major difference lies in the very small amount of clay particles in dune sands. These inorganic clay particles have a diameter of less than about 2 μm, so that in sand dunes they tend to be leached downwards with the rainwater which also washes out carbonates from the system. Leaching is a continuous process in the dunes because the sands do not overlie a bedrock which is being gradually weathered, as do most terrestrial soils, but if anything are accreting more depth of sand. Even in fixed dunes, which are covered by vegetation, the soils consist mainly of sand (quartz) particles mixed with organic humus derived from plants, but without fine clay particles. Many terrestrial soils, in contrast, contain large proportions of very small particles made up of clay minerals. These clay particles aggregate together, and also coat the larger particles, so that the soil structure consists of particle aggregates rather than assemblages of individual un-connected sand particles. This in turn means that terrestrial soils have a great variety of effective particle sizes, and also have a better capacity for water retention than sands: water is retained not just *between* the particles, but *within* the particle aggregates. When the soil dries out, water evaporates first from the external surfaces of the aggregates, and only then does it move out from the insides of the aggregates. The states of water in soils, and the factors affecting its evaporation, are clearly summarized by Vannier (1983). Overall, the water contents of terrestrial soils are not only higher but more constant than those of sand dunes, providing a more favourable environment for animal life. The diversity of animal life in terrestrial soils in turn enriches the soil by comminuting it, aerating it, and incorporating further organic matter into it.

In summary, terrestrial soils have a more complex structure than dune soils, providing an environment which is often more stable and more favourable. The critical phase for animal lines invading the land via sand ecosystems would therefore have been the initial movement into the sand dunes. Lines that penetrated these would have had few problems in taking the next movement into terrestrial soils. The rest of this chapter is therefore concerned with the possible movement of ancestral marine lines from intertidal sands to sands beyond the direct influence of the sea.

### 5.2.4    *Interstitial microfauna: ciliates and testate amoebae*

Terrestrial soils contain enormous densities of protozoa, especially ciliates and testate amoebae. Although these protozoa have a small standing biomass, their rate of production may be as great as that of the earthworms, so that their ecological importance in the soil is in no doubt. Their origins, on the other hand, are not so clear (see Foissner, 1987). It is most appropriate to discuss the ciliates and testate amoebae in turn.

The ciliates of high-shore marine sands and terrestrial soils show few features in common. Both are adapted to life in the pores between particles, but soil species are usually small and can form protective resting cysts, while marine species are large and cannot form cysts. Freshwater ciliates are also larger than soil forms, but some can form cysts, and it seems to be generally accepted that they gave rise to soil forms. This freshwater origin is discussed further in Chapter 8. There remains the possibility that *some* coastal marine ciliates produced terrestrial lines (Foissner, 1987). Most of the marine-sand ciliates are long thin 'vermiform' types, with an average length of over 400 μm, although they live in relatively small pores. Soil ciliates average only just over 100 μm in length. Other differences suggest that there are quite separate communities in marine sand, soil and fresh water. While it is conceivable that some marine forms might have given rise to species living in coastal soils, marine ciliates do not in general seem to have invaded terrestrial soils directly.

The story for testate amoebae is different, but again most authorities have concluded that there has been no direct invasion of terrestrial soils from marine sands. In this case, the forms in marine sand are smaller than those on land or in fresh water: average body lengths are 13 μm in marine sand, 72 μm in soil and 103 μm in fresh water. Some of the marine sand forms are thought to have a direct origin from the sea but the majority probably have a freshwater or even a terrestrial origin. They therefore represent the re-invasion of marine habitats, and not a primary invasion.

Both ciliates and testate amoebae have sufficient physiological adaptations to allow them to form diverse communities in supralittoral marine sands. It is not at all clear why these do not seem to have invaded the land directly.

### 5.2.5    *Interstitial meiofauna: the nematodes*

A variety of taxa of meiofauna are found within the interstitial environment of sandy beaches, including mites, tardigrades, turbellarians, gastrotrichs, ostracods and oligochaetes. Of these, the oligochaetes are discussed in Section 5.2.6. The dominant taxa, however, are nematodes and harpacticoid copepods (McLachlan, 1983), nematodes being more common in finer sediment, and copepods more common where sediments are coarse. Before considering the nematodes in more detail, it is useful to look at the overall distribution of meiofauna on sandy shores (Fig. 5.4). A comparison between two beaches in western Australia provides a good starting point (McLachlan, 1985). One of these has large accumulations of detached seaweed thrown up from offshore kelp beds, and one is a relatively clean sand beach. On the clean beach, the greatest meiofaunal densities were found about 1m down, directly below the area reached by the highest swash from the waves. Densities were not very high, however, never reaching above $500.cm^{-3}$. Meiofauna were more common on the beach with accumulated wrack, greatest densities being found at the top of the beach, just before the start of the dunes. Here there were more than 500 animals per $cm^3$, and these high densities reached from the water table, 1 m down, to the sand surface. These distributions emphasize the fact that, although the physical structure of the interstitial habitat is important in controlling animal populations, the food supply is perhaps more important still. On wrack-dominated beaches, the wrack fuels a varied community of both macrofauna and meiofauna. Macrofaunal food webs are based on those detritivores capable of breaking down marine macrophytes, particularly amphipods (see 5.2.9), but meiofaunal food webs are probably fuelled directly by the organic material actually leaching out of the wrack; this provides the primary food source, allowing the growth of bacteria, which are either eaten directly by meiofauna, or are eaten by protozoa which in turn provide food for oligochaetes and nematodes.

Food supply is therefore also likely to be a major factor when comparing the communities of sand beaches and sand dunes, although this has yet to be quantified. On clean sand, organic material is in short supply, and in this case the dunes, with their clothing of vegetation, must provide a region of sudden increase in food. The transition from wrack-covered beaches to dunes, on the other hand, is probably as abrupt in the opposite sense, with a

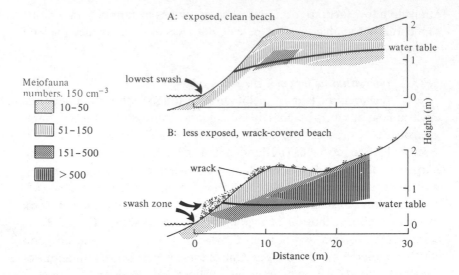

Fig. 5.4. Distribution of meiofauna on two sand beaches in western Australia. On the exposed beach there was no deposited wrack (seaweed torn from the substrate), and meiofauna numbers were low. On the less exposed beach, wrack was deposited in thick layers. Meiofauna numbers on this beach were high. The meiofauna probably depend upon bacteria for food, and these in turn are dependent upon organic leachates from the wrack. After McLachlan (1985).

sudden diminution of available organic material at the seaward edge of the dunes. Trophic interactions at the beach/dune interface now require detailed investigation.

Despite the distinct changes in food supply across the sand flat/sand dune interface, the interstitial habitat does provide a continuous link between marine sands, dune sands, and terrestrial soils. The groundwater (all water present between the sediment particles) is continuous between all these habitats, and some meiofaunal taxa, especially nematodes, are found over the whole of this span. The idea that all porous substrates form a distinct environment, allowing animals to move between major habitat types while not presenting the harsh barriers found on the surface of the substrate, has been proposed by Vannier (1983). Vannier termed this environment of pores the 'porosphere', and pointed out that many animal lines may have followed this route when moving from water to land. The term porosphere embraces similar habitats to those inhabited by animals with a cryptozoic life-style but is somewhat more specific in that cryptozoic animals also live in leaf litter and in crevices, wherever the

environment is relatively constant with high humidity and little temperature stress.

One of the problems of the porosphere of sand dunes, as a route for
nematodes from sea to land, is that the groundwater there tends to be both
scarce and variable in supply, except below the water table. Nematodes,
however, are 'aquatic' animals in the sense given in Chapter 2, and are
active only when covered by a film of water. Nevertheless, many nematodes
can survive almost completely desiccated, in a state known as 'cryptobiosis', when most life processes are immensely slowed down or
suspended. Return to a normal physiological state occurs on re-hydration,
even if this is delayed for years. Many are also thought to be able to regulate
body volume efficiently when threatened with osmotic swelling, and are in
these ways ideally adapted to an environment where water may change in
both availability and salt content (Little, 1983).

The hypothesis that nematodes may have invaded terrestrial soils via
coastal sands is supported by observations of the distribution of species in
the groundwater of regions between marine sands and terrestrial soils – the
'Kustengrundwasser' (coastal subsoil water). Here, marine nematode
species are found together with terrestrial species, and this may suggest that
not only have marine forms invaded the land, but that terrestrial lines may
have re-invaded the sea. Surprisingly few investigations have considered
the fauna of this intermediate region, however, and detailed conclusions
must await further study.

### 5.2.6    *Interstitial meiofauna: the oligochaetes*

The high densities of meiofauna on sand beaches with wrack
deposits were mentioned in the last section. Near the top of the beach, these
communities contain very large percentages of oligochaetes. The two
commonest families found in marine sands are the Tubificidae, which are
abundant in fully saturated sands low on the beach, and the Enchytraeidae,
which are dominant on the upper beach slopes, often in association with
wrack. The interacting physical and biological variables controlling their
distribution have been admirably summarized by Giere & Pfannkuche
(1982). While substrate characteristics are important – with oligochaetes
perhaps being more abundant among angular particles rather than
rounded ones – water content is probably of major importance in affecting
distribution, in combination with the availability of oxygen. Enchytraeids
in particular appear to be more tolerant of a wide range of conditions, and
this must be linked with their high-shore distribution.

Until recently, most authorities believed that oligochaetes were primitively a freshwater group (4.9.2), and that marine oligochaetes must therefore be secondarily derived. Although some authors hold to this view, believing that the similarity of organization of all oligochaetes points to a limnetic origin (Giere & Pfannkuche, 1982), others point to the fact that some groups of oligochaetes, particularly in the family Tubificidae, are more common in the sea than in fresh water (Erséus, 1984). Within this family, indeed, some subfamilies are established in the deep sea, although the majority are found in or near the littoral zone. It may be, then, that the marine tubificids evolved directly from marine ancestors, and have *not* re-invaded the sea from fresh water.

It is now possible to take this argument further. Closely related to the Tubificidae is the family Enchytraeidae. Many species of enchytraeids also are marine, and these comprise a large proportion of the interstitial forms in littoral sands. Many of these marine forms are tolerant of desiccation, and penetrate into dune sands, where moisture levels are below 1%. The majority of enchytraeids are terrestrial, living especially in soils with a very high water content (Healy & Bolger, 1984). Very few, in contrast, are characteristic of truly sublittoral *freshwater* habitats. It is at least a possibility, therefore, that the stem form giving rise to both Tubificidae and Enchytraeidae, a proto-tubificid, may have evolved in marine conditions, and that invasion of terrestrial soils by these groups occurred directly through the interstitial habitat.

If this hypothesis is correct, the picture differs from that envisaged for the origin of other terrestrial oligochaetes, the earthworms, which almost certainly moved into freshwater sediments before moving on to land (Brinkhurst, 1984; see 8.6). The consequence is that the two dominant oligochaete groups of terrestrial soils, the enchytraeids and the earth-worms, although now co-existing, may have had quite separate routes on to land. So far no comparisons between these two groups have been carried out to test this hypothesis, but a detailed consideration of the differences between them might well now prove of interest.

### 5.2.7    *The origins of terrestrial insects, myriapods and onychophorans*

Present-day marine sands contain negligible numbers of insects and myriapods, if one excepts those living inside the wrack deposits. Those that are present on upper shore sandflats, such as beetles and centipedes, are clearly of terrestrial origin, and are usually associated with the edges of the dunes, and dune vegetation. Although they demonstrate that modern myriapods and insects can adapt to marine intertidal conditions, they provide no clues about how the groups originally became terrestrial. As discussed in Section 4.9.3, many authorities, such as Manton (1977), link the origins of myriapods with those of the insects and onychophorans, and group them as the 'Uniramia'. Since many believe that they are derived from a common ancestor, they are here considered together. Nevertheless, theories about their origins are divergent and uncertain. Some authors, such as Hinton (1977) have suggested that insects might have originated on land, but the majority probably believe that they were originally derived from some aquatic ancestor perhaps via a myriapod or myriapod-like intermediate. Ghilarov (1959) was the first to suggest that the soil might have been an intermediate habitat for insects or myriapods invading the land from the water. The high osmotic pressure of body fluids suggests that they may have an origin direct from the sea rather than from fresh water (4.9.3), but this type of evidence cannot suggest which particular route they may have taken across marine shores.

Little insight into the routes of insects on to land can be gained by discussing the adaptations of pterygote insects, but a consideration of the apterytotes, and of the smaller groups of myriapods such as the Symphyla, may provide some clues to their ancient origins. In particular, the features concerned with water balance and reproductive processes suggest that many myriapods and apterygote insects are essentially adapted for a cryptozoic existence in the soil.

The symphylans are myriapods that are common in soil and leaf litter, and occasionally in the marine intertidal zone. They are very intolerant of desiccation, and probably rely mainly on behavioural mechanisms to reduce water loss: characteristic sense organs on the head can probably detect humidity changes. They do have labial excretory organs, which from their fine structure appear to be involved in producing urine while conserving salts and water, but perhaps more importantly, they have a number of small abdominal sacs that can be inflated and used for absorbing water from the substrate. Reproduction involves production of a stalked spermatophore by the male, which is attached to the substrate. The female bites off the spermatophore to obtain the sperm, a method of fertilization possible only in humid cryptozoic niches.

Of the apterygote insects, some of the Thysanura closely resemble the Symphyla in these details. The machilids are found in forest leaf litter, and in crevices at the top of marine shores, where *Petrobius* spp. are common. These also have labial excretory organs, and abdominal sacs for water uptake. Reproductive behaviour here involves the attachment of sperm droplets to a thread by the male, who also manoeuvres the female into position so that she can take the sperm up in her ovipositors. Most other apterygote insects – Diplura, Protura and Collembola – show surprisingly similar adaptations. All have abdominal sacs or their derivatives for taking up water from the substrate, and all have labial excretory organs. All these three groups, unlike the Thysanura, also have a cephalic humidity receptor which may be homologous with that of the Symphyla, and all produce spermatophores in the soil which are picked up by the females (Fig. 5.5). There are also many differences between these groups, which I have not emphasized here, but the similarities suggest in all of them a basic adaptation to the cryptozoic niche or porosphere. The use of sperm drops and spermatophores, especially, is a characteristic adaptation to interstitial habitats, and probably arose with the aquatic origin of the groups, a considerable time before they invaded terrestrial soils.

It is suggested, therefore, that primitive insects and myriapods evolved in marine interstitial habitats; either they evolved from a common ancestor, or insects evolved from myriapod forms, but both were primarily adapted for cryptozoic habitats. It is impossible to date their origins accurately, but the presence of fossil Collembola in the Devonian suggests a time earlier than this. If the Silurian fossils at present attributed to Myriapoda (1.1) are truly myriapods, the origins of the Uniramia could have occurred as long ago as the Cambrian. These forms would all have been small, probably fragile interstitial animals, so that they have not been preserved as fossils.

The subsequent emergence of pterygote insects from the cryptozoic habitat into the desiccating macro-climate did not occur until the Carboniferous, in association with the development of a variety of terrestrial vegetation. The lepismatoid Thysanura give us some indication of how this may have occurred, as they are the only apterygotes adapted to really dry habitats. Several adaptations allow species such as *Thermobia domestica* to live in deserts, or more recently in human habitations. One of these is the development of water-proof epicuticular waxes, which dramatically cut down evaporative water loss. These waxes probably evolved first in a quite different context in the cryptozoic niche: in Collembola, for instance, they are responsible for forming a hydrofuge surface, so that the animals are surrounded by an air layer even under water. This anti-drowning mechanism was later expanded to become a water-proofer. Another major adaptation was the development of a rectal system for re-absorbing water from the faeces, and in *Thermobia* this system also absorbs water vapour from unsaturated air. Rectal re-absorption has evolved in conjunction with the development of Malpighian tubules for the formation of urine, and the loss or modification of the labial kidneys. Perhaps surprisingly, reproduction in *Thermobia* involves the

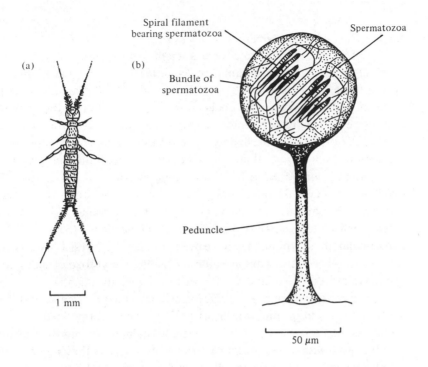

(a)

(b)

Spiral filament
bearing spermatozoa

Spermatozoa

Bundle of
spermatozoa

Peduncle

1 mm

50 μm

Fig. 5.5. An example of an apterygote insect and its spermatophore: a) shows
the dipluran *Campodea*. b) shows a spermatophore, supported off the ground
by a peduncle, and containing sperm bundles attached to spiral filaments.
Partly after Bareth (1968).

production of spermatophores, but the male produces a spermatophore
only when the female is ready to pick it up. Nevertheless, there is no internal
fertilization, which appears to have been a still later development of the
pterygotes.

It is not possible to say at what stage the ancestors of pterygote insects
and of myriapods such as the centipedes left the marine interstitial habitat.
While most of the apterygotes and symphylans just discussed have moved
on to land in soil systems, and have remained there, pterygotes, centipedes
and *Thermobia* could have emerged from the interstitial habitat at the top
of sand beaches. If the date of their evolution is as far back as postulated
above, there would at that time have been little competition from other
groups, and there would have been no major predators on land, apart from
the chelicerates, which are considered next.

### 5.2.8     *The origins of terrestrial chelicerates*

The chelicerates are a very ancient group, with a fossil record extending back to the Cambrian. One class of chelicerates – the Xiphosura or horseshoe crabs – is still found in the sea, and the Pycnogonida or sea spiders, which may also be chelicerates, are also found there. Most chelicerates – including the spiders, scorpions, and mites – are, however, terrestrial, or have secondarily moved into fresh water. The evidence from the composition of body fluids (4.6) suggests, as with the uniramians, a direct origin from the sea rather than origins via fresh water. Because the fossil record of even the terrestrial groups stretches back to the Devonian, however, their origins are not at all clear. It now seems probable that several chelicerate groups have invaded land independently (Savory, 1971). In particular, the scorpions may have been pre-adapted to resist desiccation, while most other chelicerates are adapted to life in cryptozoic niches, and may have moved on to land via the interstitial environment.

The cryptozoic forms are widespread, are mostly small, and can withstand desiccation only for short periods. These forms include some spiders, some pseudoscorpions, harvestmen, palpigrades, and the uropygids, but, even within these orders, there are often species that have become adapted to withstanding harsh environments. Some spiders, for instance, can live on the surface of sand dunes, withstanding both very high and very low temperatures. There is, then, little direct evidence about their origins, but it may be appropriate to point out their close co-evolution with the insects: chelicerates have evolved as predators of the insects, and most terrestrial ecosystems are dominated by insect-chelicerate interactions. It is likely that this trophic link is an old established one, and that chelicerates have since their origin been associated with insects. If this is the case, many chelicerates may have accompanied the early insects in their evolution in the marine interstitial environment.

The evidence concerning scorpions suggests, in contrast, that they have always been relatively large animals, suited to living for most of their time on the surface of the substrate, or burrowed within it, but certainly not inhabiting the interstitial pores and crevices of sediments. It is true that forest-dwelling scorpions live in leaf litter and beneath bark, but they are not small enough to live within soil spaces. The remote ancestors of scorpions may have been the eurypterids, which were large epibenthic animals, and the earliest fossil scorpions were themselves probably aquatic (Rolfe, 1980). Since some of the eurypterids lived in fresh water, and, since some of the early scorpion fossils occur in fluvial (freshwater) sediments, it is quite possible that some scorpions moved on to land through fresh or brackish habitats (Selden & Jeram, 1989). Indeed it seems probable that several separate scorpion lineages took different routes to land. In spite of this, however, adaptations to land have been relatively uniform, involving the replacement of external gills by internal book-lungs, and the replacement of sensory setae with longer, thinner sensory hairs for use in air, but retaining most of the primitive characters. Present-day scorpions are very widely distributed in terms of habitat, from the cryptozoic niches of tropical forests to desert sands, and it is impossible to deduce their original terrestrial habitat. Many of the larger species are burrowers, using their abilities either to form their own burrows, or to catch their insect prey (e.g. Brownell & Farley, 1979), but the smaller forms, as mentioned above, hide in crevices. At present, we cannot even guess at whether the ancestral forms utilized marine sand beaches as an environment in which to burrow, or whether they lived in the crevice habitats of rocky shores.

### 5.2.9 *Burrowers: the isopod* Tylos *and the amphipod* Talitrus

So far this chapter has been concerned almost solely with animals small enough to live between the sand grains, in the porosphere. Several groups of larger semi-terrestrial animals have made use of sandy shores by employing the burrowing habit, and three of these groups will be discussed: isopods and amphipods in this section, and decapods in Section 5.2.10.

Above the strand line on sand beaches there are often dense populations of isopods or amphipods. The isopods (family Tylidae) show a very similar lifestyle to the amphipods (family Talitridae). Both burrow in the sand in day time, and emerge at night to feed on the strand-line detritus. This behavioural pattern, controlled by inherent rhythms, enables each group of animals to avoid both the low humidities on the surface of the sand in the day, and the depredations of visual predators. Since both groups dig their burrows *above* the tide mark, it is appropriate to enquire whether their present-day terrestrial relatives might have evolved from sandy beach ancestors.

Older views suggested that, because the family Tylidae is well adapted for terrestrial respiration, the family might have re-colonized marine sands from a terrestrial habitat. Recent studies by Hoese (1983), however, have shown that within the family there is a series of species in which the respiratory organs become progressively more enclosed, and presumably better adapted to breathing air while preventing respiratory water loss. The most primitive of these could have evolved from marine ancestors. In addition, the family contains another genus, *Helleria*, which shows even further terrestrial adaptations, and is found in inland woodlands. On the present evidence, therefore, it seems reasonable to accept that the Tylidae are primitively marine in origin, and are not secondary colonizers of sand beaches.

Besides the morphological and physiological adaptations for air breathing found in the marine *Tylos* spp., behavioural adaptations have played an important part in their exploitation of the high-shore sand habitat. *Tylos punctatus*, for instance, a west-coast American species, burrows as deep as 60 cm in daytime, the burrows reaching sand with a moisture content of at least 1% (Holanov & Hendrickson, 1980). *Tylos granulatus*, a species of southern Africa, also burrows above the high tide mark, but only to a depth of about 30 cm (Marsh & Branch, 1979). Its emergence is governed by a circadian activity rhythm, which is geared to the time of nocturnal low water (Fig. 5.6). Over a series of days, the period of activity therefore becomes later each night. When the period would be so late that it would fall in daylight, it 'switches back' to the previous low tide, which by this time occurs in darkness. Activity also varies over the spring/neap cycle: at spring tides, when the driftline algae are wet, *T. granulatus* is relatively inactive. At neap tides, when the deposited algae have dried out to a more suitable consistency for consumption, activity is at a maximum.

The adaptations of *Tylos* spp. allow the family to be a dominant one in an environment where the sand surface exhibits very harsh physical characteristics. The animals live an essentially terrestrial life except that they utilize a marine food source, and may depend upon a continuous supply of salts from the sea. The possibility that their ancestors could have inhabited sand dunes seems high. At the present time tylids are not found on sand dunes, but present-day distributions have much to do with factors such as competition with groups which are more recent invaders of the land: the dominant surface detritivores on modern sand dunes are probably the beetles, and these may effectively exclude the isopods. The route across the dunes on to land was at least a possibility before such competition arose.

There are many parallels to the tylids in the life of the supralittoral amphipods, *Talitrus* spp. The best known of these, *Talitrus saltator*, burrows above high water mark, emerging at night to move down the beach and feed on the strand-line algae. Its emergence is governed by a circadian rhythm, and degree of emergence varies over the spring/neap cycle, but the time of greatest activity is not on neap tides as it is for *Tylos*. Instead, maximum emergence occurs 5–7 days after full/new moon, on declining spring tides (Williams, 1979). Minimum emergence occurs in the 5–7 days

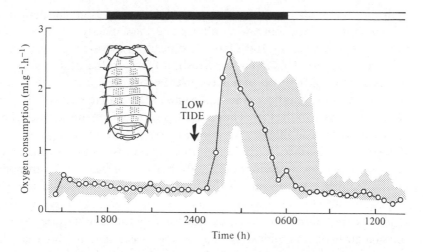

Fig. 5.6. Respiratory rhythms of the sand-beach isopod *Tylos granulatus*.
Circles and line show a typical result from one animal. Stipple shows the range
found in six animals. Horizontal black and white bar shows night and day.
The animals became active after low tide at night, and became inactive at or
after dawn. On the beach this would represent re-burrowing in the sand. After
Marsh & Branch (1979).

before the full/new moon, at a time when many of the animals are moulting
inside their burrows. Like *Tylos*, but unlike the saltmarsh amphipods,
*Talitrus* shows no circa-tidal element in its behavioural rhythms. It does
not burrow in advance of tidal inundation, but moves up the sand in front
of the advancing tide (Williams, 1982). Like *Tylos*, then, *Talitrus* has a
relatively terrestrial, if nocturnal, existence. Yet it is not the talitrids from
sandy-shores, but those from saltmarshes that are thought to have
produced terrestrial lines (4.7.2). This route is considered further in Section
6.1.5. The failure of sandy-shore talitrids to produce terrestrial lines is even
more surprising when contrasted with the success of the tylids, and requires
some discussion.

First, it is important to point out that talitrids in saltmarshes had the
possibility of moving directly into the cover provided by terrestrial
vegetation, with a reliable water supply, while on sandy beaches the
transition involved an area of little vegetation and low and erratic water
supply. The movement on to land of saltmarsh amphipods such as those
related to *Orchestia* spp. was therefore an easier step than that faced by
inhabitants of sandy beaches such as *Talitrus* and *Tylos*. Secondly, it must
be noted that both *Talitrus* and *Tylos* are admirably adapted to existence

high up on sand beaches, with excellent orientation mechanisms, behavioural rhythms, burrowing ability and the capacity to brood their young in brood pouches. It is therefore difficult to ascribe reasons for one line becoming better adapted to terrestrial life than the other. There is, however, one striking difference between the two groups, and this lies in the adaptations for respiration. Amphipod gills are large plates attached to the coxae, at the base of the legs. They are not protected from desiccation, and none have ever been shown to develop internal cavities that might protect the respiratory surfaces from desiccation. In contrast, the gills of isopods are flat plates at the posterior of the body, partly protected by the uropods, and in many species, including the tylids, they have developed internal respiratory structures known as 'pseudotracheae' (see Chapter 9). While it is always speculative to assign causes for evolutionary 'failure' to one physiological or anatomical system, it does seem that in this case the respiratory pre-adaptations of the isopods were superior to those of the amphipods in terms of preventing water loss, and that because of this the amphipods have been restricted to the 'easier' routes on to land. Their subsequent lack of physiological adaptation to various facets of terrestrial life may reflect this mode of access to land. The ability of isopods to exist in almost all habitats (4.7.1), may reflect, in contrast, their evolution in more 'difficult' habitats such as sandflats, as discussed here, and on shingle and rocky shores (see Chapter 7).

### 5.2.10    Burrowers: the decapod Ocypode

In the crab family Ocypodidae, the best-known genus is probably *Uca*, comprising the fiddler crabs. Many species of *Uca* are found on sandy shores, but, since the saltmarsh species of *Uca* will be discussed in Section 6.1.4, the present section will deal with the genus *Ocypode*, comprising the ghost crabs, which are confined to sand.

Ghost crabs are so called because they are a pale sandy colour, and are well camouflaged against their sand background, when stationary. They can move at high speed, however – *Ocypode ceratophthalma* can run at more than $2 \text{ m.s}^{-1}$ (Burrows & Hoyle, 1973) – and while running they can change direction sharply to avoid capture. They have excellent vision, with eyes raised vertically above the carapace to give a better field of view. Some species are said to be able to detect movement at distances over 100 m. The wide field of these eyes also allows them the advantages of binocular vision.

Extreme mobility and acute visual sensitivity are utilized by *Ocypode* in its feeding habits. *Ocypode ceratophthalma*, for example, has several modes of feeding (Hughes, 1966). It uses sand particle feeding, sorting organic fragments from the substrate as does *Uca*, but also is a scavenger and a predator. On the east African coast, this species preys on a smaller ocypodid, *Macrophthalmus*, either perceiving it visually and chasing it, or probing in the sand with tactile sensors and then digging it up. It also feeds on a bivalve, *Donax*, which it locates by probing in the sand with its chelae.

To an even more extreme degree than the isopods and amphipods discussed in Section 5.2.9, *Ocypode* spp. are much too large to be able to inhabit interstitial spaces. *O. ceratophthalma*, for instance, has a carapace width up to 2.5 cm. The crabs dig burrows in the sand at and above the high water mark, and these are relatively permanent structures compared with the temporary burrows of *Tylos* and *Talitrus*. As in the burrows of the high-shore *Uca* species, these burrows are a dominant feature in the life of ghost crabs (Hughes, 1966). Individuals may dig several kinds of burrows, depending essentially upon the state of the tide, but often use a burrow as a centre from which to make foraging expeditions. In some situations an individual will retain a burrow for a period of days, while in others new burrows are dug regularly. In both cases, however, the crabs spend the majority of their time in the burrows, which provide protection both from predation and from physical stress. The burrows are regarded as home territory, and an area around each is aggressively defended.

Overall, the ghost crabs appear well adapted to the habitat at the top of sandy shores, being active only in air, and taking refuge in the burrow at high tide. Their visual acuity and rate of movement are comparable with those of the terrestrial insects. It is tempting to associate these with a nearly terrestrial existence, and to suggest that their complex behaviour patterns are further adapted to an aerial life than are those of species living lower on the shore. In fact, however, we know so little of the visual and behavioural adaptations of lower shore species such as *Scopimera* and *Dotilla* that such generalizations are premature. Certainly *Scopimera inflata*, which lives in the upper half of east Australian beaches, and *Dotilla fenestrata*, which lives in the lower half of east African shores, both have rhythmic adaptations to the periods of tidal emersion, dig burrows for protection and have good vision (Hartnoll, 1973), so that they resemble *Ocypode* spp. in many ways. They also have complex adaptations allowing them to breathe air, which involve thin 'windows' in segments near the base of the legs (Maitland, 1986), and are essentially only active when not covered by water. Whether they are really less adapted to an aerial existence than *Ocypode* spp. is yet to be shown.

Finally, it should be pointed out that no ocypodids are fully terrestrial, and that all species live close to the top of the tide mark or below it.

Behavioural and physiological capability seem unlikely to be blocks to a movement on to land, but as with several other groups of crabs, they have retained a marine pelagic larva, and this forms an essential tie with the sea.

## 5.3    Conclusions

Looking at the surface of a modern sandy beach, it is hard to imagine that this might be the route by which several invertebrate lines have moved on to land. The top of the beach and the neighbouring dunes present an inhospitable environment during daytime, with wide temperature fluctuations and an erratic water supply. Yet it is concluded in this chapter that several animal lines – nematodes, oligochaetes, myriapods/insects, chelicerates and some isopods – have not only invaded this habitat but have moved further inland to become wholly terrestrial. This is in distinct contrast to the low importance of mudflats in the terrestrial invasion. The initial reasons for animal lines moving into the high intertidal with its 'difficult' environment may have to do with the rich food source often provided by the drift line. After this initial invasion, animals adapted to the conditions there would have been well adapted to move inland, where the physical environment is, if anything, less exacting. In part, then, the successful movement of animal lines on to land through sand systems can be said to be *because of* the harsh regimes at the top of the shore. In circumstances where the intertidal zone grades directly into terrestrial vegetation, in contrast, animals may move on to land, but they are not likely to have adopted such rigorous physiological adaptations for combatting water loss and temperature change, and these animals are consequently limited to very humid, sheltered terrestrial niches.

In particular, I have suggested that some of the interstitial forms have used the 'porosphere' as a route on to land, because only in this habitat are the conditions sufficiently protected from external variations in climate to allow the growth of communities of very small animals with their large surface area: volume ratios. It is because the porosphere is better developed in sands than in muds that sandflats rather than mudflats are envisaged as being an important stage in the invasion of land. The suggestion that the interstitial habitat of sands provided conditions for the origin of many terrestrial arthropod groups is more speculative than similar suggestions for oligochaetes and nematodes, but such a hypothesis can now be tested to some extent by analysing in detail the adaptations of the primitive insects and myriapods to the porosphere.

Many groups that are represented in interstitial waters have not been considered here – copepods, ostracods and turbellarians, for example.

Some of these have probably moved on to land from the interstitial waters of freshwater habitats, and are considered in Chapter 8. Nevertheless, these and other groups, including macrofaunal taxa, have in the past formed part of the communities on or within sand beaches, and interactions between them have certainly occurred. These mutual interactions provide a major complication in what has been presented here as an apparently simple story of the conquest of new physical habitats. For example, invasion of the sand habitat by proto-insects was probably accompanied by an invasion of predatory chelicerates, so that the two groups co-evolved from the start of their terrestrial existence. Other predatory groups such as the turbellarians and some nematodes have probably also played a great part in determining the community structure in sand beaches, although the relative importance of predation and competition within the meiofauna has been little investigated. The effects of the macrofauna on the meiofauna of sand beaches is also unknown, but could be important. For example, macro-faunal predators such as the ocypodid crabs may have both direct effects on their prey, and indirect effects on substrate characteristics.

In the past, these biological interactions would have been determined, at any one time, by the degree to which various taxa had invaded the intertidal sand habitat. Little evidence is available about the timing of these invasions for the smaller animals because of their poor preservation. Larger animals such as the crabs appear to have arrived on land relatively late, however (Fig. 1.1), so that they were probably not important in affecting early meiofaunal communities. Many of these communities are also found deep within the sediment, so that interactions within the community may have been more important than influences from outside. When the factors controlling present-day interstitial communities are better understood, it may be possible to carry this line of conjecture further.

# 6

## Saltmarshes and mangrove swamps as routes on to land

Have you ever been in a mangrove swamp, my dear – Nature at its *most* revolting . . . if you can imagine the smelliest forest merely growing in a *deep* drain, and, of course, the *most* emetical *red* and *black* crabs quite clambering about the branches, my dear, too nephritic.
*Punch*, 1929.

The introduction to Chapter 5 pointed out that the upper parts of sheltered shores are usually colonized by vegetation. In temperate zones this comprises mostly non-woody vascular plants, and is termed saltmarsh. In the tropics, the majority of the area between mean high water of neap tides and mean high water of spring tides is colonized by trees or sea grasses, although small areas of saltmarsh exist. The salt-tolerant trees form a characteristic community known as mangrove swamp. Both they and the saltmarsh may have provided important crossing points for animals invading land. As discussed later, distinctions between saltmarsh and mangrove plants may not have held in the past, but for convenience the two are considered separately here.

### 6.1    Saltmarshes

#### 6.1.1    Physical and biological influences in saltmarshes

The change from mudflat to saltmarsh is often a sudden one. Very often there is a 'saltmarsh cliff', anything up to 1 m high, forming the front edge of the marsh. Here physical conditions change from the open unprotected flats to a region in which vegetation offers physical protection from desiccation, temperature changes, currents, and predation, and provides a direct source of food. At the same time, some aspects of saltmarshes present harsher conditions than those on the flats: because saltmarshes reach as high as mean high water springs, and then merge into terrestrial habitats, the periods of tidal cover are shorter, less frequent, and

116

less predictable than lower on the shore (Ranwell, 1972). Salinity is therefore wildly variable. Biological influences also differ from those on the flats: only a few wading birds feed on saltmarshes, but land-based predators such as racoons and owls are common (see, e.g., Teal, 1962).

The fauna and flora of saltmarshes vary throughout the world, and conditions in the marshes therefore also vary widely. The most well-studied marshes are those of the east coast of north America and those of Europe, and it is worth contrasting these two types of marsh. An excellent account of the ecology of east-coast saltmarshes in north America has been provided by Teal (1962) and Teal & Teal (1969). In these marshes the conditions governing animal life are to a great extent determined by two grass species, *Spartina alterniflora* and *S. patens*. *S. alterniflora* is the primary colonizer of sand- and mudflats, and can live as far downshore as mean tide level. Once established, the plants decrease the speed of current flowing over the flats, thus promoting the deposition of sediment. This sediment is then stabilized by the growing *Spartina*, and the marsh level may rise rapidly. The rise in level means that less tides cover the later stages of a growing marsh, and in this later phase *S. alterniflora* is replaced by *S. patens*. While *S. alterniflora* grows to a height of 2 m, but dies down in winter, *S. patens* grows to only about 60 cm, but retains a mat of dead leaves in winter. The two species therefore provide different degrees of cover for marsh animals throughout the annual cycle.

During the summer period of growth, the *Spartina* leaves ensure that the marsh surface remains at a relatively even temperature in comparison with the neighbouring mudflats, much as in the situation described for a grass meadow in Section 2.4.3. In the winter the *S. patens* marsh continues to insulate the mud surface, but in the lower *S. alterniflora* marsh, the mud surface is free of its protective covering. In spite of the protective influence of the two *Spartina* spp., water evaporates from the soil, and to a greater extent at the very top of the marsh, where it is seldom covered with sea water. In this upper marsh, the salinity of the interstitial water can therefore rise dramatically in summer. Because rainfall is more likely to wash salt out of this top marsh, it can also be greatly diluted, and the overall salinity regime is extreme, varying from nearly fresh water to maybe twice the salinity of sea water. Although the marsh therefore *appears* to merge gradually into truly terrestrial vegetation, there is in reality a 'barrier' region at the top of the marsh in which conditions for animal life are in some ways more difficult than those in *either* the lower marsh, *or* on land. The physiological 'difficulties' of living at the top fringe of saltmarshes are emphasized by the adaptations of plant species. Although most saltmarsh plants are perennials, plants at the top of the marsh are mainly annuals, because conditions there are too harsh for them to survive over the entire year (A. J. Gray, 1985). The barrier zone found on saltmarshes is a feature common to most types of shore, and it may be the conditions within this zone that are most critical for potential invaders of land.

Conditions in European saltmarshes are in many cases similar to those on parts of the north American marshes (Long & Mason, 1983), and particularly where the main floral constituent is the recently evolved *Spartina anglica*. While American marshes reach as low down as mean tide level, however, *S. anglica* is not usually found below mean high water neaps, so that European marshes usually cover less area than their American counterparts. The rate of growth of *S. anglica* marshes can be very high: vertical accretion rates of sediment of up to 15 cm/year have been recorded, so that marshes reach maturity in a very short time. As the marshes increase in height, the time for which they are submerged decreases, so that they change from 'submergence marshes' to 'emergence marshes'. A critical point may be reached when they are continuously out of water for 10 days or more (Ranwell, 1972). In this 'emergence marsh', *S. anglica* grows taller than in lower regions, offering more shelter to marsh animals. In European marshes, then, taller grasses are to be found at the top of the marsh, in contrast to American marshes where the tall grass is found at lower levels.

Other types of saltmarsh are also common in Europe, particularly those consisting of what is known as the 'general saltmarsh community' – a mixture of plants in the middle levels of the marsh such as *Limonium, Armeria, Puccinellia* and other species in a mosaic pattern (Chapman, 1960; Adam, 1981). These species do not die back in winter, and the saltmarsh provides shelter for invertebrates throughout the year. In one marsh on the east coast of England, for instance, Marsden (1976) has shown that, while daily fluctuations in the temperature of the air above the vegetation may be as much as 30 °C in summer, the maximum temperature change at the base of the vegetation is only 4 °C. Indeed, of all the habitats on the marsh, including salt pans, the soil on the marsh surface was the most thermally stable.

The mixture of plant species provides a mosaic of different micro-habitats on the saltmarsh, in much the same way as has been shown to occur on rocky shores (Chapter 7). In addition to this small-scale variation in conditions, larger scale variation is provided by the drainage creeks which have better-drained soil and whose banks are dominated by *Halimione*. On the surface of the marsh there are also often large numbers of 'saltmarsh pans' – small depressions of very variable shape where water is retained on the marsh. In these pans a specialized aquatic fauna lives, in close proximity to the more terrestrial species of the saltmarsh surface.

Present-day marshes in the northern parts of the temperate zone have only a short history, having originated after the last glaciation some 10,000 years ago. The first saltmarshes, however, probably originated in the Cretaceous, soon after the origin of the angiosperms themselves. Whether there was any equivalent habitat on sheltered shores before the emergence of the angiosperms is very doubtful. Certainly, any such habitat would have been quite different if it was covered by algae. A variety of present-day algae *live* on saltmarshes, but they cannot be said to form them. Some, such as *Bostrychia* and *Pelvetia* are unattached and depend upon the sheltered conditions between the angiosperm stems to stay in place. Others such as *Vaucheria* are effective in binding mud surfaces and producing accretion, but provide little shelter for other than very small animals. There are no algal associations that produce any equivalent to either *Spartina* marshes or the mixed saltmarsh community. In the past, however, groups other than the angiosperms and the algae may have provided saltwater communities. The 'Rhyniophyta' are thought to have lived at the edge of the sea in Silurian times (Chapter 1), and later plants may have produced equivalents to the modern saltmarsh. Although the true angiosperm saltmarsh as an intermediate for faunal lines invading the land can therefore only have been of importance for the later groups – those becoming terrestrial in or after the Cretaceous – there may have been opportunities for early animals to move directly from sea to land through other types of vegetation.

Angiosperm saltmarshes are also important in the re-colonization of semi-marine habitats by terrestrial animals, particularly spiders and insects. This is presumably because the angiosperms provide an environment and a food source more similar to terrestrial conditions than any algal community. Saltmarsh faunas of marine origin consist primarily of molluscs, such as the prosobranch *Littoraria irrorata* and the pulmonate *Melampus bidentatus*; crustaceans such as the talitrid amphipods; and the ocypodid and grapsid crabs *Uca* and *Sesarma*. These and the numerous coleopteran, dipteran and hemipteran insects will now be discussed in turn.

### 6.1.2 *Pulmonate gastropods:* **Melampus** *and* **Ovatella**

In saltmarshes on the east coast of north America, the dominant snail is *Melampus bidentatus*, a pulmonate of the family Ellobiidae. In the upper parts of the marshes, where *Spartina patens* and short growths of *S. alterniflora* predominate, there may be 1300 snails per square metre (Price,

1980; Price, 1984). The snails are active in the summer periods, but burrow or hide in crab burrows during the winter. Because the zone occupied by *Melampus* is covered by water for only 2–4% of the time, the problems of desiccation can be extreme, and some individuals have been shown to lose 40% of their body water in the field. In the laboratory, the snails can tolerate a loss of up to 80% of their body water, and can re-hydrate rapidly afterwards. The processes of osmotic 'tolerance' discussed in Chapter 4 are evidently paramount here.

Like all pulmonate snails, *Melampus* breathes through a vascularized mantle cavity or lung, and has no gill. It appears to be well adapted to breathing in air, as its uptake rate when in air is about 10 times that when under water. Although it can tolerate several days continuous immersion in water, depending upon the temperature and salinity involved, it drowns eventually because it cannot use the lung as a 'physical gill'.

Physiological adaptation to extreme conditions is far from being the sole adaptation of *Melampus* to life at high tidal levels. The reproductive cycle is strictly geared to the frequency of inundation by spring tides, so that the critical phases in the cycle occur at two-weekly intervals in summer (Russell-Hunter *et al.*, 1972). The snails aggregate when the high spring tides cover the top of the marsh, then copulate and lay their eggs. Hatching occurs two weeks later when the next set of spring tides covers the marsh. From the benthic egg masses hatch veliger larvae, which spend approximately two weeks in the plankton. These planktonic larvae then settle back on to the marsh a further two weeks later, again coinciding with coverage by spring tides (Fig. 6.1).

In Europe, the commonest saltmarsh pulmonate is *Ovatella myosotis*. Like *Melampus*, it lives high up on the marsh, sometimes beneath a mat of *Puccinellia*, and sometimes at the extreme edge of the marsh in the detritus beneath the shrubby species *Suaeda fruticosa*. *Ovatella* shows many physiological parallels with *Melampus*, including an extraordinary ability to tolerate changes in salinity. This species can withstand a change in the salt concentration of the mud on which it lives from freshwater to a salinity of nearly three times sea water (Seelemann, 1968). Its reproductive habits are different, however, in that *Ovatella* lays gelatinous egg masses from which hatch not pelagic larvae but miniature snails. As far as is known, there is no strict link between the various phases of reproduction and the incidence of high spring tides.

As outlined in Section 4.8.1, interest in these members of the Ellobiidae has been heightened by the suggestion that they represent a state similar to that in the primitive ancestors of the true land snails, the Stylommatophora

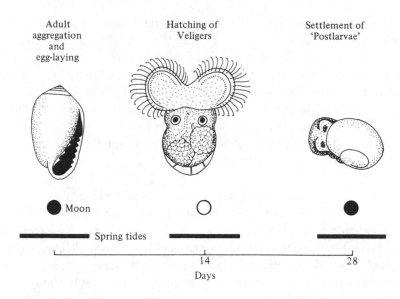

Fig. 6.1. The reproductive cycle of the saltmarsh pulmonate *Melampus bidentatus*. Adults aggregate when high spring tides cover the marsh, then lay eggs on the marsh surface. At the next set of high spring tides, veliger larvae hatch and begin planktonic life. These veligers settle back on to the marsh as 'postlarvae' when the next set of high spring tides covers the marsh surface. After Russell-Hunter, Apley & Hunter (1972).

(Morton, 1955). Some of the Ellobiidae have themselves become terrestrial: the genus *Pythia* contains large forms found in forest bordering the coastline in the Indo-Pacific, and the genus *Carychium* contains minute snails found in inland beechwoods in Europe. The most primitive of the ellobiids, however, are those found in saltmarshes, such as *Ovatella* and *Melampus*. There is no doubt that these species demonstrate the type of physiological and behavioural pre-adaptation necessary for possible invaders of the land. They have successfully become adapted to the harsh 'barrier zone' at the top of the saltmarsh, and little more than behavioural changes would be needed to adapt them to a terrestrial existence. Whether they were themselves the actual forerunners of the Stylommatophora has, however, recently been questioned (Solem & Yochelson, 1979). One of the major problems is the discovery that there was a diverse fauna of terrestrial pulmonates in the Carboniferous, with four of the present-day families already represented. As yet there is no evidence that the Ellobiidae evolved before the Jurassic, some 150 million years later. Unfortunately, however, the fossil record of freshwater and terrestrial molluscan faunas is exceedingly sporadic, and it is hard to be sure that ellobiids did not exist earlier than the Jurassic. Even if the origin of terrestrial stylommatophorans from ellobiid-like forms did occur in the Carboniferous, this early date means that the intermediate habitats were certainly not saltmarshes as we know them, made of angiosperms. Although the physiological evidence about the origins of osmoregulatory tolerances points to a direct marine or estuarine origin for the terrestrial pulmonates, ideas about their origins in salt-marshes must remain analogies.

### 6.1.3    *Prosobranch gastropods:* Littoraria *and* Assiminea

In the same marshes containing the pulmonate *Melampus*, the commonest prosobranch gastropod in the lower zone of *Spartina alterniflora* is *Littoraria* (formerly *Littorina*) *irrorata*. This snail, like most other marsh snails, is a detritivore. It feeds on the marsh surface at low tide, but as the tide rises it retreats up the plant stems. It then draws the body into the shell, and seals the lip of the shell on to the stem with a mucous thread. It migrates down again as the tide falls. Despite belonging to a family of typically marine snails, then, this species is almost entirely active out of water.

The adaptive significance of the movement up the *Spartina* stems probably concerns the avoidance of predation by decapods such as *Callinectes sapidus* (Fig. 6.2), which forages for prey under water, and can have a devasting effect upon *Littoraria* populations when they are experimentally held down at the marsh level (Warren, 1985). Snails held on the marsh also experienced more osmotic problems than those allowed to climb plant stems: they were submerged in pools during rainstorms and became bloated with absorbed water. In the normal situation, snails congregate together at the base of plant stems during hot days, minimizing water loss; and their plant-climbing activity may be beneficial in terms of temperature control, because gastropods that reduce their surface contact with the substrate have been shown to stay cooler than those that attach to the substrate with the foot (Vermeij, 1971). Together with *L. irrorata*'s high temperature tolerance (it can withstand 40°C), this behaviour allows the species to survive over a high temperature range.

*Littoraria irrorata* orientates visually to the stems of *Spartina* (Hamilton, 1978). When placed on bare sand or mud the snails crawl directly towards *Spartina* stems or artificial substitutes during daylight, without apparently using any sun-compass orientation mechanism. The eyes, situated on short protuberances near the base of the tentacles, have nearly spherical lenses

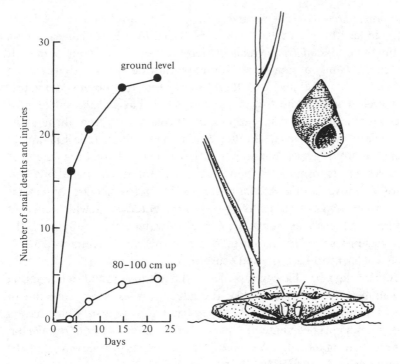

Fig. 6.2. The effect of height above the saltmarsh surface on injury to the snail *Littoraria irrorata*. Snails on the marsh surface suffer heavy mortality, mainly from the crab *Callinectes sapidus*. Snails tethered high up to *Spartina* stems escape predation and show very little injury. After Warren, 1985.

and can focus from 2.3 mm to infinity when in air. They are larger than in species such as *Littorina littorea*, which is primarily active in water, have more photoreceptors, and therefore presumably have better resolution (Hamilton *et al.*, 1983). As will be discussed in Chapter 9, one of the trends in evolution from water to air is the switch from primarily chemical cues to primarily visual cues.

Further species of *Littorina* will be discussed in Section 7.4. They are particularly important for the understanding of the invasion of land by molluscs because within the same superfamily, the Littorinacea, there is an extensive terrestrial family, the Pomatiasidae. These terrestrial snails were briefly discussed in Section 4.4. They appear to have evolved directly from marine ancestors, and not to have had any intermediate freshwater stage. The date of origin of this terrestrial family is thought to be in the Cretaceous, so that here at least is a possible contender for an origin in saltmarshes.

*Littoraria irrorata* is a relatively large snail, adults reaching up to 20 mm in length. A contrast in size is provided by a European saltmarsh prosobranch, *Assiminea grayana*, which reaches only about 5 mm. It belongs to a family of small snails found worldwide in brackish water and saltmarshes, the Assimineidae. *A. grayana* lives at the top of saltmarshes, and therefore spends most of the time in air. It has no gill, unlike most aquatic prosobranchs, and presumably takes up oxygen through its vascularized mantle cavity (Fretter & Graham, 1962). It is tolerant of violent salinity regimes, but shows little capacity to osmoregulate, and resembles in its osmotic responses the pulmonate *Ovatella myosotis* discussed above (Little & Andrews, 1977). In particular, it remains substantially hyper-osmotic to the environment over a wide range of salinities, and this may allow it to take up water osmotically from the substrate. In spite of living out of water, it therefore maintains a water supply sufficient to keep up body turgor.

*Assiminea* lays benthic egg capsules in the saltmarsh mud in late spring and summer. These capsules, like the adults, are extremely tolerant of changes in environmental salinity. They hatch into veligers which have a free-swimming pelagic phase. In this *Assiminea* resembles not *Ovatella* but the American pulmonate *Melampus*. Nothing is known, however, of the timing of veliger hatching in *Assiminea*.

The terrestrial relatives of *Assiminea* have not been examined in detail. One species of *Pseudocyclotus* lives on the leaves of bushes in Papua New Guinea, an unusual habitat for terrestrial prosobranchs – most are found in leaf litter or at least on the ground. In this arboreal habitat it is certainly subjected both to extreme desiccation and also to dilution by rainstorms. The origins of this and other terrestrial assimineids are not known for certain, but evolution in the harsh environment at the top of saltmarshes would have provided ideal pre-adaptations for an arboreal life.

### 6.1.4    *Crabs: grapsids,* Sesarma *spp., and ocypodids,* Uca *spp.*

American saltmarshes support a wide variety of crabs from several families. The life-styles of the grapsids *Sesarma cinereum* and *S. reticulatum*, and the ocypodids *Uca pugnax* and *U. pugilator* serve to illustrate the various ways in which crabs are adapted to saltmarsh life (Teal, 1959; Teal & Teal, 1969).

*S. reticulatum* lives in the lower, muddier parts of the marsh, and is active when covered by the tide, or when the weather is cloudy and there is little chance of desiccation. Each individual has a burrow in the marsh, to which it can retreat, and each burrow has several openings so that when the tide rises it rapidly fills with water. *S. cinereum* provides a striking contrast. It lives at high levels on the marsh and is active only in air. It has no burrows like those of its low-shore relative, and when the tide rises it climbs up the vegetation and remains above water level. When these two crabs were investigated in the laboratory, their different reactions to water cover were indicated by their rates of respiration. *S. reticulatum* had a higher rate of oxygen uptake under water than in air, probably because of the increased activity needed to ventilate the gills under water. *S. cinereum*, on the other hand, responded to water by remaining inactive, so that it had the same rate of oxygen uptake in air as in water.

It is not known whether the differing adaptations of these two crabs to different zones in the marsh are purely behavioural, or whether they also show differences in structural and physiological adaptation. They have similar rates of water loss through the integument when in air, so the high-shore *S. cinereum* has no obvious advantage over *S. reticulatum* in terms of desiccation resistance. On the other hand, the rate of water loss in air is much lower than that from low-shore xanthid crabs, and higher than that of nearly terrestrial ocypodids (Herreid, 1969). *Sesarma* spp. therefore show a moderate degree of adaptation to semi-terrestrial life, albeit mainly in terms of behaviour.

The two fiddler crabs, *Uca pugilator* and *U. pugnax*, have been investigated more thoroughly, but even so their major adaptations to life in the marsh appear to be behavioural. *U. pugnax* lives in the middle marsh and is mainly active when the tide is out, but may continue foraging while covered by water. It digs a burrow with only one entrance, and seldom blocks this. *U. pugilator* lives on the high marsh, and is active solely when out of water. As the tide rises, it retreats to its burrow and blocks the entrance with a plug of mud or sand. It therefore remains in a pocket of air for most of the high tide period. The exclusion of water from the burrow is, however, probably not concerned with respiratory adaptations. This species *does* have a higher rate of oxygen uptake in air than in water, and can therefore be regarded as better adapted to aerial than to aquatic respiration; but the same is true of *U. pugnax*, which does *not* maintain air in its burrow at high tide. Both species, in fact, have their gill chamber partly adapted as a 'lung' and breathe air. The reason for the difference in the use of plugs is more to do with maintaining the integrity of the burrow. In *U. pugnax*, the burrow is made in firm mud along the roots of *Spartina alterniflora*, and there is no tendency for collapse under water; but in *U. pugilator*, the burrow is usually in sandy mud with only sparse growth of *Spartina patens*, and, if water penetrates into it, the structure collapses.

The importance of the burrow structure can best be understood in terms of its use in reproduction. Males defend their burrows, and females ready to mate follow them into the burrow. Mating takes place inside, in each case after the male has plugged the entrance. Unlike the situation in simple retreat into the burrow, then, *both* species block the burrow and maintain the most stable conditions available for mating. These conditions are maintained for incubation, since the females remain in the burrows until the eggs are about to hatch. The egg masses are large, and protrude from under the abdominal flaps, so that they are very vulnerable to crushing if the burrow collapses (Christy & Salmon, 1984). Many of *U. pugilator*'s burrows are dug above the limit of the tides, and it is these burrows that are most in demand by the female crabs. This emphasizes the fact that the burrow has a further function: protection from predators. Particularly in the zone at the top of the marsh, there are many terrestrial-based predators such as racoons and birds. Plugging the burrow is a further protection against these.

The burrows of the two species also serve as a refuge from physical stress. Retreating into the burrow has been shown to be the most effective way of reducing body temperature, although other mechanisms such as changing body colour and allowing evaporation of water from the body surface are also useful (Smith & Miller, 1973). The problem of desiccation is also lessened by retreating into the burrow. Both species are in fact good at regulating the composition of the blood when faced with either high or low salinity (Wright *et al.*, 1984), but loss of more than about 20% of the body water is lethal to most fiddler crabs. For both species of *Uca*, therefore, the life-style is very much organized round the burrow, and much of the time spent above ground is used in defending the burrow or courting females to entice them into the burrow. Because of their relatively high tidal position, both species can spend long periods in these activities, during which they interact with other individuals. It may be because of these long emersion times that prolonged and complicated interactive behaviour patterns have developed in fiddler crabs. The enlarged claw of the male is used in aggressive displays and in courtship, and each species has specific patterns of claw-waving. Visual displays are partly replaced by stridulating at night, and again each species has specific sound patterns. It has been suggested for several groups of animals that behaviour has become more complex as the groups have become more terrestrial. This point will be taken further in Chapter 9.

While adult fiddler crabs live in saltmarshes or in mangrove swamps, the larval forms are planktonic. Ovigerous females with embryos about to hatch emerge from their burrows at night, at the time of high spring tides, and walk to the water's edge. The zoea larvae are deposited in the water, and they then live in the plankton, metamorphosing into megalopa larvae before settling back on to the marsh as juveniles (DeCoursey, 1983). The initial migration of the females to the water's edge is governed by an endogenous rhythm, and the process is reminiscent of that described for the saltmarsh gastropod *Melampus*. The larvae of *Uca* spp. are more sensitive to salinity and temperature changes and extremes than the adults (Vernberg, 1984), and, although some acclimation to temperature occurs over the seasons, the zoea larvae cannot withstand low salinities. It may be, therefore, that the pelagic larval stage in fiddler crabs has been retained because development in sea water offers a more constant environment than could be provided even by the behavioural adaptations of the adults. Many semi-terrestrial crustaceans have retained a pelagic stage, and this means that they are tied to coastal and marine environments, even when the adults are physiologically well adapted to terrestrial conditions.

The adaptations of *Uca* spp. to high tidal levels in saltmarsh habitats suggest, overall, that the saltmarsh route could have been important in allowing decapods to become terrestrial. The earliest fossil terrestrial decapods come from the Tertiary, so that, as with the littorinacean gastropods, the timing of their emergence fits well with the timing of angiosperm evolution: decapods could have become adapted to salt-marshes in the Cretaceous, and moved on to land in the Tertiary.

### 6.1.5    Amphipods: the Talitridae

Amphipod crustaceans are abundant in most marine intertidal habitats. Their adaptations to sandy shores have been described in Section 5.2.9, where it was concluded that sandhoppers have not given rise to terrestrial forms. Here the possible evolution of the landhoppers – so abundant in tropical and south-temperate rain forests – from saltmarsh forms is considered.

Saltmarshes throughout the world provide habitats for two types of amphipods: low on the marsh and in saltmarsh pans the family Gam-maridae is usually predominant, but higher up is found the family Talitridae (Daiber, 1982). The *Gammarus* species are essentially aquatic animals, moving by swimming or by wriggling on their sides. The talitrids, such as *Orchestia* spp., are in contrast essentially active in air, and walk upright over the marsh surface. It is the members of the Talitridae, therefore, that are of most interest in the context of the invasion of land.

The physiological ecology of this invasion has been discussed in detail by Spicer, Moore & Taylor (1987).

On saltmarshes in North America, the predominant talitrid is *Orchestia grillus* (Kneib, 1982), in New Zealand, *Orchestia chiliensis* (Marsden, 1980), and in Europe, *Orchestia gammarellus* (Daiber, 1982). Little experimental work has been carried out to investigate the factors determining the distribution of these species on the marshes, but for *O. grillus* it now seems likely that the preference for specific physical conditions maintains the population high up the shore: animals that were moved downshore in artificial islands of *Spartina* wrack either moved out or died (Kneib, 1982). The greatest densities of saltmarsh talitrids occur beneath strand-line detritus. On most marshes this consists in the main of decaying *Spartina*, but *O. gammarellus* is also common under the decaying wrack found on rocky and shingle shores. Talitrids are detritivores, and the strand-line wrack provides an abundant food supply.

The conditions under the wrack are very different from those on the saltmarsh surface, and provide milder and more constant physical conditions. For wrack on rocky shores, Moore & Francis (1985) have contrasted relative humidity and temperature at the wrack surface with values 15 cm down (Fig. 6.3), and have shown the relative constancy of conditions within the wrack pile. *Orchestia* spp. *do* move out from their refuges beneath the wrack, but usually only at night (Wildish, 1970). This activity pattern is governed by an endogenous circadian rhythm, and in *O. gammarellus* ensures that the animals shelter from low external humidity and from visual predators during the day, but emerge when humidity is high at night.

The behaviour of these littoral talitrids thus ensures that they are not, for the most part, exposed to extremes of temperature and desiccation. Nevertheless, as has been pointed out in earlier sections, conditions high on the saltmarsh, even under piles of wrack, are more extreme than those in the lower littoral regions, and it is appropriate to consider what structural and physiological adaptations for a semi-terrestrial life have evolved within the Talitridae.

One of the major factors that has allowed amphipods to become independent of cover by water is their method of reproduction. Instead of liberating free-swimming larvae, like most decapods, amphipods retain their young in a brood chamber, until they emerge as miniature adults. This pre-adaptation allows the developing young to mature in conditions determined by the adults, so that they avoid the rigours of desiccation in the macro-climate. The ability of adults to choose appropriate micro-climates is therefore important for succeeding generations.

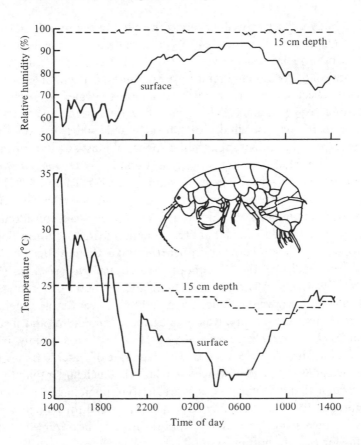

Fig. 6.3. Diurnal variations in temperature and humidity during summer in a wrack pile inhabited by the amphipod *Orchestia gammarellus*. Although conditions on the surface were very variable, the climate 15 cm down was stable. After Moore & Francis, 1985.

The physiological responses of adults to humidity and temperature extremes have been most thoroughly studied in *Orchestia gammarellus* (e.g. Taylor & Spicer, 1986; Moore & Francis, 1985; Morritt, 1987). Water loss from *O. gammarellus* is rapid in desiccating conditions, and adults may lose more than 25% of their body weight in 10 hours at 75% R.H. Relative humidities of more than 80% are required for long-term survival. When kept in humid conditions, though, *O. gammarellus* can osmoregulate well over a wide range of substrate salinities. On agar substrates, which provide food as well as support, it maintains its blood hyper-osmotic below $25^o/_{oo}$, and above this it regulates hypo-osmotically.

Respiratory gas exchange in amphipods probably occurs across the general body surface as well as across the gills. In semi-terrestrial forms, such as *O. gammarellus*, the area of gills is reduced when compared with aquatic forms such as *Gammarus* spp. (Moore & Taylor, 1984), and the gills are more solid. As in many semi-terrestrial lines, it is possible that these changes reflect a need to regulate evaporative water loss while still permitting oxygen uptake. There has, however, been no concomitant increase in the oxygen affinity of talitrid blood, which has only a low concentration of haemocyanin, similar to that of most aquatic crustaceans (Taylor & Spicer, 1986). Rates of oxygen uptake by *O. gammarellus* are approximately the same in air as in water (Spicer & Taylor, 1987).

ical adaptation to terrestrial life, and it must be concluded that behavioural adaptations are important. This point is reinforced when considering the truly terrestrial landhoppers that are a common feature of forest floors in the tropics and the southern hemisphere (Friend & Richardson, 1986). One of the best studied of these landhoppers is *Arcitalitrus sylvaticus* (Lazo-Wasem, 1984), an Australian leaf-litter species now also common in California. When given a choice of humidities it spent most time above 80% R.H. Its survival time was high at these humidities, and in this it resembles *O. gammarellus*. Below 80% R.H., it lost water rapidly, indeed, more rapidly than *O. gammarellus*. This high rate of water loss may relate to the size of its gills, two of which are relatively much larger than those of *Orchestia*, so that the total gill area is greater. The point has been confirmed by Morritt (1987) for *Arcitalitrus dorrieni*. The trend for reduction in gill size with increasing terrestriality appears here to be reversed, and the reasons for this are not understood.

It is useful at this point to consider the possible phyletic relationships of some of the genera within the family Talitridae. According to Bousfield (1984), the terrestrial talitrids have at least two separate origins. One, as represented by *Arcitalitrus*, is related to 'palustral talitrids', primitive forms living mainly in estuarine saltmarshes and mangroves. The other, represented by *Talitroides*, is related to the 'beachfleas', such as *Orchestia*, and more indirectly to the 'sandhoppers' such as *Talitrus*. When studying the evolution of anatomical and physiological adaptations to land, the most enlightening comparisons may therefore be made between the palustral talitrids and *Arcitalitrus*, and between *Orchestia* and *Talitroides*. It is not particularly surprising that there seems to be no direct trend in the series *Orchestia* to *Arcitalitrus*, since the two are only indirectly related.

Evidence from the biogeography of talitrids has allowed a reconstruction of their palaeohistory (Bousfield, 1984). It seems likely that the semi-terrestrial and terrestrial groups mostly originated in the Cretaceous, and that these origins were stimulated by the origins of the angiosperms which created abundant niches for semi-terrestrial detritivores. The prime routes to land at that time were probably the saltmarsh and mangrove environments, and the importance of these to talitrids is emphasized by the large number of species resident in present-day marine environments dominated by angiosperms. It may be concluded that the majority of terrestrial talitrid lines invaded land via vegetated habitats.

The two major lines of the invasion of land have been discussed by Spicer, Moore & Taylor (1987) and Morritt (1988). The line leading to *Arcitalitrus* and its relatives can never have been subjected to great desiccation pressure. This line may have evolved from the marine littoral zone of Gondwanaland directly into forest leaf litter. The group related to *Orchestia*, on the other hand, may have emerged from saltmarshes and mangrove swamps. The terrestrial species of this group have so far been little studied, but, if the line moved through the relatively harsh environment at the top of saltmarshes, it would be expected that terrestrial species would show stronger adaptations to life on land than the *Arcitalitrus* line. This hypothesis now needs testing experimentally.

### 6.1.6    Re-invasion from the land: insects and spiders

Saltmarsh faunas contain large numbers of insects and spiders, but these come from terrestrial, and not from marine stock. The discussion of these groups is therefore not directly relevant to the main theme of this book, but may provide some insight into the factors that control the composition of saltmarsh communities.

Insects are much less common in the marine intertidal zone than they are in fresh water or on land, as discussed by Little (1983). Within the intertidal, they reach their greatest numbers of species in saltmarshes, and this is probably due to two factors. First, the evolution of insects has been closely linked with that of the angiosperms; and of all the marine habitats, those with most angiosperms are undoubtedly saltmarshes (in contrast to mangrove swamps, in which much of the plant structures grow above high tide level). Secondly, aquatic insects are dependent to a great degree on access to the air/water interface, and in most intertidal regions this is a place of high stress from tidal movement and wave action. The most shelter from these stresses in the marine intertidal zone is found in saltmarshes and mangrove swamps, which, because of their situation and their dominant angiosperm cover, provide relatively sheltered niches.

Most saltmarsh insects belong to the Diptera, Coleoptera and Hemiptera. Of these, many of the flies are transitory, but the beetles tend to be burrowing forms and the bugs are phytophagous and closely associated with plants. Both groups can therefore find shelter from the forces of surface tension as the tide rises and falls, and they are important members of the saltmarsh foodwebs.

Spiders are often the dominant arthropod predators on saltmarshes (Long & Mason, 1983), but numbers of species are often higher on estuarine marshes and near the high tidal levels, where salinity is lower and the incidence of tidal cover is least. Like the insects, saltmarsh spiders have remained essentially air-breathing, and, although they tolerate submersion, they are active only when out of water.

## 6.2     The importance of saltmarsh as a route on to land

In Section 5.1.5, it was concluded that, although mudflat animals show many of the adaptations achieved by various groups in coping with semi-terrestrial conditions, it is not likely that mudflats have provided a major route on to land. Saltmarshes, in contrast, because they offer a mixture of marine and terrestrial conditions, and abut directly on to land, have potentially provided a direct exit route from the sea. They also provide a 'testing ground' for would-be terrestrial invaders, at the top of the marsh, where conditions oscillate violently in salinity, temperature and humidity, and where terrestrial as well as aquatic predators are forces to be reckoned with. On the other hand, saltmarshes as such are relatively recent phenomena, and cannot have existed before the Cretaceous. Many animal lines – perhaps most of them – produced terrestrial offshoots well before this (Chapter 1), and it may be that they utilized saltwater communities consisting of entirely different plant taxa. In particular, the distinction currently made between saltmarshes and mangrove swamps may have been less clear, and many of the suggestions made here for 'saltmarshes' could equally well have applied to 'mangroves' (see 6.3).

For some groups, such as the pulmonate molluscs, the importance of saltmarshes as routes on to land remains unclear. For other groups that are thought to have a more recent terrestrial origin, such as the littorinacean prosobranch gastropods, some lines of decapod crustaceans, and the amphipod crustaceans, the hypothesis that saltmarshes formed an intermediate habitat for lines invading the land seems very plausible: the timing was right, the conditions were right, and, as will be seen in succeeding chapters, many other habitats provide less likely intermediates.

## 6.3     The mangrove habitat

The last section has shown that, despite their recent origin, saltmarshes may have been important as sites in which terrestrial animal lines originated. The same may be said for mangrove swamps, which in many senses are the tropical equivalents of saltmarshes. This section will first consider the general ecology of mangrove swamps, and will then go on to discuss the adaptations shown by some groups of animals living in them.

Mangroves have an essentially tropical distribution, but in many places they spread outside the actual limits of the tropics of Cancer and Capricorn. A better general limit is perhaps the 30° latitude, but even to this there are several exceptions: mangroves occur as far south as New Zealand, and as far north as Japan. This distribution is governed by temperature, and, although some species tolerate a minimum of 10°C, most flourish only where the temperature in the coldest month does not fall below 20°C (Chapman, 1977a). Like saltmarshes, they occur at the top of sheltered sedimentary shores, often in estuaries, and usually stretching from the level of mean high water springs down to just above mean tide level. The trees grow best where rainfall is high and not seasonal, and in these areas the mangrove forest reaches heights of 30 m or more (Macnae, 1968).

The majority of mangroves are placed in four families of trees. Although there are many genera and species, the dominant genera are *Rhizophora, Ceriops, Bruguiera* (family Rhizophoraceae), *Avicennia* (family Avicenniaceae), *Sonneratia* (Sonneratiaceae) and *Laguncularia* (Combretaceae). Each genus has a characteristic form which to a great extent influences the physical conditions within a given area of mangrove swamp. In particular, two major types of root systems exist. In *Rhizophora*, the main trunk is supported by arching aerial prop roots, which in turn are anchored in the mud by radiating anchor roots. *Rhizophora* swamps are therefore dominated by tangled masses of these prop roots. In *Avicennia* and *Sonneratia*, in contrast, there are no prop roots, and from the central trunk cable roots radiate out some 20–50 cm below ground level. These cable roots bear aerial roots or pneumatophores which rise above the ground surface and are probably important in ensuring an air supply to the whole root system. Since the soils of mangrove swamps tend to be fine-grained and anaerobic, the pneumatophores are essential oxygenating organs.

Within the mangrove swamp, conditions are quite different from those on the adjacent mudflats (see, e.g. Hutchings & Recher, 1982). Since there is a great variation between different mangrove swamps, and especially between those of the Indo-West Pacific and the Atlantic, it may be most instructive to consider conditions within several particular swamps. The Port Royal swamps in Jamaica have been described by Warner (1969). Here, the seaward part of the swamp is formed by *Rhizophora mangle*, but towards the back of the swamp this species is for the most part replaced by

*Laguncularia racemosa* and *Avicennia nitida*. The *Rhizophora* zone is relatively open, so that air circulates freely, and temperatures do not rise as much as in the *Avicennia* zone, where growth is tangled and air circulation is restricted. Soil composition changes from the seaward edge, where it is clean peat, to the landward zone, where the peat is mixed with mud, sand and gravel. Port Royal has a seasonal rainfall, so that in the dry season the back of the swamp dries out and becomes hyper-saline, while in the wet season it becomes waterlogged and hypo-saline. Salinity changes at the seaward edge are minimal because salinity there is mainly governed directly by sea water. The landward edge of the swamp gives way to dry sandy soil with cactus and thorn scrub. In this swamp, therefore, most conditions become more variable and harsher from the seaward edge to the back of the swamp, but there is no gradual transition to the nearby terrestrial zone.

The Pandan mangrove swamps in Malaya have been described by Berry (1964). Here the seaward edge of the swamp is again dominated by *Rhizophora* spp., which are replaced by *Bruguiera* spp. further inland. Rainfall in this region is high, and is not seasonal, and the landward margins of the swamp merge directly into terrestrial rainforest. These mangrove swamps therefore provide a direct route on to land even for cryptozoic animals intolerant of desiccation and temperature changes.

Many mangrove swamps do not have a seaward fringe of *Rhizophora* spp. On more open shores neighbouring Pandan, the outer regions are colonized by *Avicennia* spp. (Chapman, 1977b), and the same is true on the east coast of Africa (Macnae & Kalk, 1962). Here the *Rhizophora* spp. are restricted to the edges of creeks and channels. At the climatic limits of mangroves, such as in south-east Australia and New Zealand, the diversity of mangrove species is decreased, and again *Avicennia* spp. often predominate.

Mangrove swamps provide a variety of different habitats for animal life. Prime among these are the trunks, branches, prop roots, pneumatophores and leaves of the mangrove trees themselves, which offer the rare combination of a hard surface in a very sheltered environment. Both sessile species such as barnacles and bivalve molluscs, and mobile species such as gastropods and crabs, are abundant on these hard surfaces. Within the tree canopy they are joined by animals of terrestrial origin such as insects and spiders. Of these arboreal species, some that live at low levels are filter feeders, and these are found especially near the seaward edge of the swamp. Most, however, probably feed on deposited detritus and on epiphytic algae coating the mangrove trunks. Only the animals of terrestrial origin, especially lepidopteran larvae, beetles and monkeys, eat living leaves of the various mangrove species (Onuf *et al.*, 1977; Hutchings & Recher, 1982).

On the mud surface, most food webs are based on the rain of dead leaves from the trees, although a great deal of this organic material is usually lost from the swamps, triggering high rates of secondary production on the neighbouring mudflats (Wells, 1984). Some animals, such as the pulmonate and prosobranch snails, take in leaf material that has been well broken down and mixed with inorganic mud; but others, such as many of the sesarmid crabs, drag whole leaves down into their burrows. A study by Robertson (1986) on the feeding of *Sesarma messa* in north-eastern Australia showed that this crab removed more than a quarter of the annual leaf litter fall in *Rhizophora* forest. Crab species are often the most abundant detritivores resident in mangrove swamps. Most of them form burrows, and in so doing they may alter the surface topography of the swamp. *Heloecius cordiformis*, an ocypodid crab found in the *Avicennia* swamps of east Australia, has been shown to produce well-drained mounds in the swamp, with more sand than the neighbouring muddy flats (Warren & Underwood, 1986). This probably occurs because the crabs feed by collecting balls of sediment from the flats and taking them down their burrows at low tide.

In some mangrove swamps, such as those of Port Royal in Jamaica (Warner, 1969), the burrows of various species of crab are so numerous that they anastomose, producing a network of 'crab-runs', inhabited both by the burrowers and by species that do not themselves excavate burrows. As in the crabs living at the top of saltmarshes (6.1.4), life for many mangrove crabs centres around their burrows.

While the tide is out, invertebrates such as the crabs and molluscs dominate the mud surface, together with the gobioid fishes known as mudskippers. At high tide, however, all these animals disappear into burrows, and the mangrove swamp is invaded by a variety of fish and prawns, which come in on the rising tide to feed. In a study of four fish species captured in a mangrove swamp in north-east Australia, Beumer (1978) showed that some, like the black bream *Acanthopagrus berda*, migrated into the swamp, others such as the toadfish *Chelonodon patoca* were semi-resident, while the goby *Ctenogobius criniger* was a permanent resident. All these species were fairly catholic feeders, but most fed primarily on benthic meiofauna. In the Indo-Pacific, predatory fishes may be accompanied by water snakes, which attack crabs hiding in their burrows. The influence of these aquatic predators suggests that burrowing, and possibly the arboreal habits of mangrove invertebrates, are related to predation pressure as much as to the particular physical characteristics of the environment.

To consider the adaptations of animals to life in mangrove swamps, four examples will now be discussed in turn.

### 6.3.1    *Prosobranch gastropods: the genus* Littoraria

The dominant snails living on mangrove trees belong to the prosobranch family Littorinidae. These snails were formerly known as *Littorina scabra* in the Indo-Pacific, and *L. angulifera* in the Atlantic. Recent work by Reid (1985, 1986) has shown that in the Indo-Pacific there are actually 20 species, while in the Atlantic there is only *L. angulifera*. The genus is now referred to as *Littoraria*.

*Littoraria* spp. are thin-shelled littorinids, living on trunks and leaves of mangrove trees, and usually remaining above the water level. Some show tidal migrations up and down the mangrove stems, moving up just ahead of the tide, and later moving down on to the damp bark (Nielsen, 1976). They therefore breathe air all the time, and not water, although their gill leaflets are only slightly reduced in size in comparison with aquatic littorinids. Some have been found as high as 5 m above the swamp surface, while others remain quite near the ground. They are active only in damp conditions, and at least one species, *L. luteola*, has been shown to be activated as much by terrestrial influences such as rain and dew as by tidal cycles (Little & Stirling, 1985, as *L. scabra*). In dry conditions, *Littoraria* spp. withdraw into their shells and seal themselves to the substrate with a mucous thread (McMahon & Britton, 1985).

The 20 species of *Littoraria* in the Indo-Pacific show two major types of reproductive strategy. These are characterized by differences in the female reproductive tract, and the associated processes of releasing the fertilized eggs. Some, such as *L. melanostoma*, are oviparous and have large albumen and capsule glands. These give successive coatings to the fertilized egg as it passes down from the ovary, and eventually each egg is released within a gelatinous matrix as a pelagic egg capsule. Others, such as *L. luteola*, and *L. angulifera* of Caribbean mangrove swamps, are ovoviviparous and have small albumen and capsule glands. In their case, the egg is retained in between the gill leaflets until the embryo has developed into a veliger larva. Brooding may last for only 4 days in tropical forms such as *L. angulifera*, but is prolonged to about 17 days in temperate species such as *L. luteola*. The length of time for which veliger larvae are planktonic is not known, either for oviparous or ovoviviparous species, but periods of 8–10 weeks have been suggested. Since all species have planktonic veligers, the distribution of adults must be to a great extent determined by the settling behaviour of these larvae.

The Indo-Pacific species have a variety of distribution patterns within mangrove swamps (Fig. 6.4). Many are commonest at the seaward fringe of

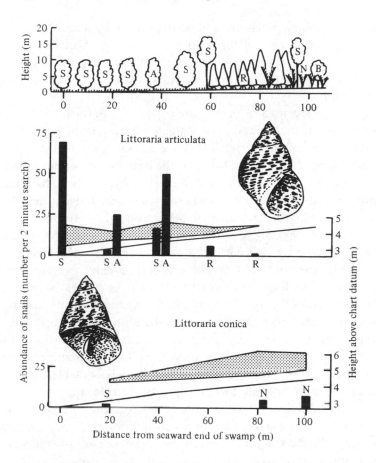

Fig. 6.4. Distribution of two *Littoraria* spp. in a mangrove swamp in Borneo. The top diagram shows distribution of mangrove species. S shows *Sonneratia*; A shows *Avicennia*; R shows *Rhizophora* and B shows *Bruguiera*. N shows the palm *Nypa*.

The second two diagrams show the distribution of *L. articulata* and *L. conica*. Letters under the columns indicate the tree species occupied. The stipple shows vertical distribution. *L. articulata* inhabits predominantly *Sonneratia* and *Avicennia* at the front of the swamp, while *L. conica* occupies mainly *Nypa* at the back of the swamp. After Reid (1985).

the swamp, on *Avicennia*, which is presumably the first settling place reached by the larvae on the incoming tide. They are not so common on *Rhizophora*, and this may be partly because this genus has hard shiny leaves, whereas *Avicennia* has hairs on the undersides of its leaves, and these are grazed by the snails. The diet of *Littoraria* spp. has not been investigated in detail, however, and it is assumed that most species gain a living by consuming diatoms and perhaps other algae on the bark of the trees. The distribution of the species may also be affected by predation pressure, as mentioned above. For instance the arboreal crab *Metopograpsus* is a major predator of *Littoraria*, and its distribution may influence that of the snails.

A very few species of *Littoraria* are commonest at the landward edge of the swamp, and these are perhaps of greatest interest from the point of view of the origins of terrestrial littorinaceans. Of the tropical species, *L. carinifera* and *L. conica* are found at the back of the swamps, mainly on *Avicennia*, but also on the mangrove palm, *Nypa*. Their distribution is centred in Indonesia, in areas where rainfall is high and constant. In these regions, the mangrove swamp often borders directly on tropical rainforest. Since *Littoraria* spp. are to a great extent governed by terrestrial influences such as precipitation, it is relevant to ask what prevents them from moving into purely terrestrial habitats. The answer is probably twofold. First, and most obvious, both species produce planktonic egg masses. No present-day *Littoraria* species lays benthic eggs, nor do any brood their young past the veliger stage, although this brooding is found in the European littorinid, *Littorina saxatilis*. Secondly, the snails probably depend to some extent upon a supply of salts from sea water or from mangrove leaves. No information about their osmoregulatory abilities is available, but it is likely that they cannot tolerate total independence from a sea water supply. Nevertheless, these species can be seen to have evolved a near-terrestrial life-style. It is quite likely that an ancestral littorinid living in mangrove swamps, which laid large benthic egg masses and became tolerant of limited salt supplies, could have moved directly into the terrestrial ecosystem of tropical forests. Some scenario such as this can be envisaged for the ancestry of the Pomatiasidae, the terrestrial littorinaceans. The possibility of their origin from saltmarsh ancestors was discussed in Section 6.1.3, and the close relationship of saltmarsh to mangrove habitats is emphasized by considering a final species, *L. luteola*. This species is found along the whole length of the eastern coast of Australia. In the temperate regions, as far south as 37° S, it is found from the seaward edge of the *Avicennia* swamps right to their landward fringe. It is a catholic species in terms of habitat. It occurs on both leaves and trunks of *Avicennia*, is found on the small mangrove shrub *Aegiceras*, lives on fallen logs as well as living on mangroves, and also inhabits saltmarsh vegetation to the landward of the swamps. Since one species can inhabit such a variety of habitats, it does not seem impossible to conceive of an ancestral relative making the transition to an environment independent of the sea.

### 6.3.2    *Prosobranch gastropods: the genus* Nerita

Accompanying *Littoraria* on the mangrove trunks, but at a lower level, are found prosobranchs of a different family, the Neritidae. Like *Littoraria, Nerita* spp. can be found from the front to the rear of the swamps (Berry, 1964). They spend most of their time out of water, and, although they retain a gill, they are essentially air-breathers (Houlihan, 1979). The mantle floor is well vascularized, and they probably take in oxygen as much through this area as through the gills. Neritids have internal fertilization and lay benthic egg masses on the mangrove trunks, but like *Littoraria* these hatch into planktonic veliger larvae. Their reproductive cycles are governed more by terrestrial climates than by tidal cycles (Berry, Lim & Sasekumar, 1973).

From these parallels with *Littoraria*, it might be thought that marine neritid lines could have given rise directly to terrestrial neritaceans. The evidence, however, suggests that this is not so, but that *Nerita* has followed the mangrove route into fresh water (Macnae, 1968). Subsequent emergence from fresh water may *then* have produced terrestrial forms (see 8.5). This freshwater route has been possible because many mangrove species are tolerant of reduced salinity. *Avicennia marina*, for instance, can grow in water that is almost fresh, and some species of *Sonneratia* and *Bruguiera* are restricted to waters of less than $10°/_{oo}$ S. According to Chapman (1977a), many mangroves can flourish in fresh water when competition is reduced. Neritids appear to have followed this invasion of fresh water, and truly freshwater species are common worldwide.

### 6.3.3    *The grapsid crab* Aratus pisoni

The dominant invertebrates on the swamp surface are crabs. A whole variety of ocypodids, grapsids and gecarcinids are active at low tide and retreat to their burrows on the flood, as described in Section 6.3. Most of these crabs are more or less restricted to the mud substrate, but a few, such as *Metopograpsus* in the Indo-Pacific and *Aratus* in the Atlantic, are truly arboreal, and are thus perhaps more intimately associated with the mangrove trees themselves. Many of the adaptations of the mud-dwelling crabs are similar to those of crabs in saltmarshes, and those of the gecarcinids are considered in Section 6.3.4. Here it is appropriate to follow one arboreal grapsid in detail.

In the Port Royal swamps of Jamaica, Warner (1967, 1969) described the distribution and life history of the grapsid *Aratus pisoni*. This is most common at the seaward edge of the swamp, reaching $16.m^{-2}$, but is also frequent right across the swamp to the landward side. It is active during the daytime at low tide, and is the commonest crab in this particular swamp. *Aratus* is lightly built, and adults spend much of their time in the mangrove trees. They are susceptible to desiccation, however, and have to return to the sea to re-hydrate. Females also migrate to the seaward fringe of the swamp to release their larvae when they hatch from the egg masses attached to the pleopods. Hatching occurs in phase with lunar cycles, occurring at both new and full moons (i.e. at the times of spring tides), usually at night. The female walks down a *Rhizophora* root to the water, enters it head first, and then vibrates the abdomen rapidly. This releases the prezoeae in a cloud into the water, and the female then re-emerges. Because hatching occurs *before* the female enters the water, she spends only a minimal time there, vulnerable to aquatic predators.

Larval *Aratus* probably spend about a month in the plankton, and then settle in the swamp as juvenile crabs. These are more widely dispersed across the swamp than the adults because they live in the communal 'crab-runs' beneath the mud surface and are not subject to the problems of desiccation found at tree level. Adults and juveniles probably also avoid predation by the large crab *Goniopsis cruentata* by their behaviour: adults being out of reach in the trees, and juveniles hiding in the crab-runs.

There are very few grapsid crabs which live inland, away from the coastal zone. Those that do so are found in tropical rain forest associated with epiphytic bromeliads which retain sufficient fresh water in their leaf axils to provide a water supply all the year round. Their origins are thought to have been *via* freshwater systems (Little, 1983). The mangrove grapsids such as *Aratus* have therefore apparently failed to produce truly terrestrial lines. The reasons for this are not understood, but it seems probable that reproductive mechanisms have been important. With few exceptions (see Chapter 8), grapsid crabs have retained marine pelagic larvae, and even when the adults have moved inland, they migrate back to the coast to release their zoea larvae. Although the retention of pelagic larvae suggests a parallel with the marine prosobranch gastropods, some snails have become entirely emancipated from the sea by laying benthic egg masses with sufficient nutriment to allow development to the juvenile forms without external water supply other than that available in the soil. This procedure has not been adopted by any grapsids, nor indeed by any decapods.

### 6.3.4    *Gecarcinid crabs:* Cardisoma *and* Gecarcinus

In company with grapsids and ocypodids, many gecarcinid crabs are typical of mangrove swamps. *Cardisoma guanhumi* is abundant in mangrove swamps on the east coast of Central and South America, the Caribbean and Florida. It is a burrowing species, and, although it may reach 500 g in weight, densities of nearly 2.m$^{-2}$ have been recorded in favourable areas (Herreid & Gifford, 1963). The major populations are found in mangrove swamps, but *C. guanhumi* is also found in low-lying areas up to 5 km from the sea. Its burrows reach down from 30 cm to 1.5 m below ground level, and terminate in a chamber which is always partly filled with groundwater. Near the sea this may be essentially sea water, but inland the water may be completely fresh. The crab needs to moisten its gills in this water, and if deprived of access to it will usually die within a few days.

In Florida, *C. guanhumi* has a peak spawning period in October and November, individual spawning phases occurring at new and full moons. The females carry the egg mass, and migrate to the sea as this ripens. The zoea larvae are released into the sea in much the same way as for *Aratus*, but in the case of *Cardisoma* the breeding migration may involve a journey of up to 5 km (Gifford, 1962). The species also shows mass migrations, involving both males and females, and which are presumably concerned with a dispersal mechanism, although the basis for these movements is not understood.

A close relative of *C. guanhumi* is *Gecarcinus lateralis*, also found in the Caribbean region. *G. lateralis* can also be found near the sea, but it usually occurs further inland. It digs burrows in grassy areas where there is a heavy dewfall, but these burrows do not reach down to groundwater like those of *Cardisoma* (Fig. 6.5). It may be able to absorb water from dew by using setae which project down externally from posterior ventral openings of the gill chamber (Bliss, 1979). It cannot use these setae to suck moisture from interstitial sand spaces, as can some rather more aquatic ocypodid species (9.1.1) and even, to some extent, *Cardisoma* (Wolcott, 1984). Within the burrow, however, it is possible that it can absorb water osmotically from damp soils: the crab presses itself close to the substrate, and, because of the difference in osmotic pressure between its body fluids and the interstitial soil water, water moves into the animal by osmosis (Wolcott, 1984). This method can function with less available water than the physical suction method, and may be typical of more terrestrial animals.

Reproductive strategies of *Gecarcinus* are similar to those of *Cardisoma*, in that the female carries the developing egg mass attached to her pleopods, and migrates to the sea to release the eggs by vibrating her abdomen. The hatching larva is a planktonic zoea, and this spends some time at sea before the young crabs settle back on the shore.

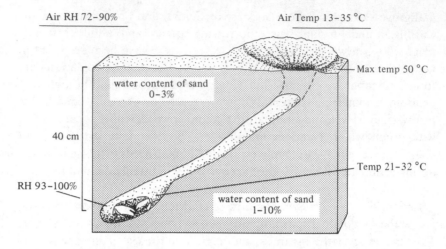

Fig. 6.5. Conditions in and around the burrow of the land crab *Gecarcinus lateralis*. The burrow is a refuge from extreme climatic conditions in the air and on the sand surface. After Bliss (1979).

Unlike the grapsids, the gecarcinids can be seen to have moved inland through the intermediate habitat of mangrove swamps. Although they are still linked to the sea by their retention of a pelagic larva, their burrow-digging habits have allowed them to become independent of a supply of liquid water. As emphasized by Wolcott (1984), *G. lateralis* survives more by adaptations which allow conservation of water supplies than by mechanisms which allow it to take up liquid water. Since liquid water may be scarce on land, the advantage of relying on conservation and slow uptake within the burrow can be appreciated. Certainly, *G. lateralis* is one of the most 'terrestrial' of known crabs, although its relative inability to prevent evaporative water loss from the integument has meant that it is usually restricted to activity at night except in humid micro-climates (Bliss, 1979). The factors that limit *Gecarcinus* to the maritime region have been discussed by Wolcott & Wolcott (1988). Contrary to expectations, there seems little reason to suppose that lack of salts is a limiting factor. *Gecarcinus* can survive with a purely terrestrial salt supply, and its inability to penetrate far inland remains as yet unexplained.

### 6.3.5 *The cryptozoic assemblage: the nemertines* Pantinonemertes *and* Geonemertes

So far the smaller, soft-bodied members of the mangrove fauna have been mentioned only as food items for the macrofauna. There are, however, numbers of polychaetes and sipunculans, and smaller animals such as nematodes among the benthic fauna of mangrove swamps (Hutchings & Recher, 1982). Among the soft-bodied animals there are also carnivorous forms, the nemertines, and these will be taken as an example of the cryptozoic fauna.

While gastropods and crustaceans utilize the external surface of mangrove trees, the nemertine *Pantinonemertes winsori* lives beneath the bark, and in cavities below it, in rotting mangrove trunks in north-east Australia (Moore & Gibson, 1981). This habitat has a high humidity, but is seldom covered by water, and the nemertines crawl out of sea water when submerged in the laboratory. They can move slowly by ciliary gliding on a mucous track, but also have the capacity for rapid locomotion by extending the massive proboscis. The tip of this is attached to the substrate, and the body is pulled forward by contraction of the proboscis itself. The proboscis is presumably used mainly for prey capture, as it has large, sharp stylets at its end.

Although *P. winsori* has only been described from mangrove swamps, a related species, *P. californiensis*, is found at the top of saltmarshes on the west coast of north America (Gibson, Moore & Crandall, 1982). This species occurs underneath rotting logs where the tides reach only infrequently, and its existence emphasizes the strong link between mangrove swamps and saltmarshes. Like *P. winsori, P. californiensis* is active in humid air, but will not survive if kept under water.

A third species in the same genus as the above two species is *P. agricola.* This species lives in Bermuda, and has been recorded not only from mangrove swamps but from the lower intertidal *and* from fully terrestrial situations, in moist earth under stones. In other Caribbean and Indo-Pacific islands, further, related, species occur. These are placed in the genus *Geonemertes*, and are all fully terrestrial.

The occurrence of terrestrial species which are apparently closely related, but are widely separated geographically, may suggest that independent invasion of the land has occurred several times from one basic marine stock. The use of mangroves as an intermediate habitat for these several different invasions seems likely, especially when species such as *P. agricola* can be found both within mangrove swamps and on land. Further discussion will be given to the route by which a separate group of nemertines may have invaded land in Section 7.9.

### 6.3.6    *Mudskippers:* Periophthalmus *and* Boleophthalmus

Among marine teleost fish, the family Gobiidae contains the largest number of present-day species. Several genera, including *Perioph-thalmus* and *Boleophthalmus*, are characteristic inhabitants of mangrove swamps. *Boleophthalmus boddaerti* is usually found at the seaward fringes of mangroves in the Indo-west Pacific, and extends on to the soft mud beyond the trees (Macnae, 1968). It therefore spends the majority of its time in air, and has somewhat reduced gills, together with some vascularization of the buccal cavity. Further into the swamps are found species of the genus *Periophthalmus*. Some of these, like *Boleophthalmus*, seem to remain submerged when the tide rises over the swamp, but others such as *P. chrysospilos* actively avoid being covered by water, and climb into the mangrove trees. They are able to do this because the pectoral fins of mudskippers lie far forward, and are modified into suckers. The distribution of a number of species in a north-east Australian mangrove swamp, and their interactions, have been described by Nursall (1981).

While some mudskippers, like *B. boddaerti*, sift through mud and extract algae, many species are predators, taking small invertebrates from the mud surface. Some species follow the tide up into the swamp as it rises, preying upon both marine animals and insects. They have good vision, provided by eyes set on the top of the head, and they can move rapidly by 'skipping'. This movement entails the rapid straightening of the tail from a flexed position, and results in a leap on to the prey. Mudskippers can also move more slowly by levering themselves along, using their pectoral fins as crutches, and supporting their weight alternately on these and on the pelvic fins.

For adult mudskippers, life is centred round a burrow (Fig. 6.6), as is the case for many crabs. Here, however, the burrow is used mainly for breeding. In *P. chrysospilos*, for instance, the burrow is constructed by the male in soft mud. It then fills with water from the neighbouring flats. The male attracts a female by a display involving the erection of the dorsal fins and inflation of the orange-coloured throat. Once pairing is established, a territory around the nest burrow is vigorously defended. Eggs are laid on the inside walls of the burrow, and hatch as typical gobioid larvae (Macnae, 1968).

The life of mudskippers is clearly amphibious. They are active on the surface of the mangrove swamps at low tide, but are also able to use the water-filled nest, and, indeed, to swim well in water. The physiological processes that have been investigated fit them well for this lifestyle. They can absorb oxygen from both air and water. They can move efficiently in air and when submerged. In water, they excrete ammonia, but when in air they can store their waste nitrogen as urea until they return to water (Little, 1983). Despite these adaptations, though, it is believed that mudskippers have not given rise to any terrestrial lines. Terrestrial vertebrates are

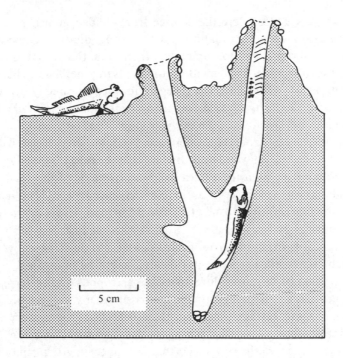

Fig. 6.6. Section of a breeding nest of the mudskipper *Periophthalmus sobrinus* in East Africa. The 'chimneys' of the nest are made of mud pellets deposited by the fish, and several pellets can be seen at the base of the burrow. The marks on the side of the taller tunnel are made by the pectoral fins as the fish climb upwards. After Stebbins & Kalk (1961).

thought to have evolved exclusively from freshwater, and not marine stocks. This direct parallel with neritid snails cannot be explained unequivocally, but limitations in reproductive strategy are suggestive. As will be discussed in Chapter 8, one of the major adaptations of the vertebrate lines that have produced terrestrial offshoots from fresh water was the development of the embryo to a relatively late stage before hatching. This development has in the main been restricted to freshwater vertebrates, while marine vertebrates, like marine neritids and grapsid crabs, have remained tied to the sea by a planktonic larva. Reproductive adaptations have obviously played a large part in directing the invasion of land, and will be discussed in more detail in Chapter 9.

### 6.4 Conclusions: angiosperm-dominated habitats as routes to land

In regions where rainfall is seasonal, there is usually a restrictive barrier to the invasion of the land at the landward edge of both saltmarshes and mangrove swamps. Here the change from shaded, moist, relatively cool habitats to open, dry habitats with strong diurnal temperature variations is abrupt, and for animals other than those that burrow, probably overwhelming. Movement on to land is in general more likely to succeed in areas where intertidal vegetation leads directly to thick terrestrial cover. In the tropics, this occurs in regions where rainfall is high and constant, and where the mangrove swamps abut on to tropical rain forest. In the temperate zone, equivalent areas occur where saltmarsh abuts directly on to terrestrial woodland. In this chapter it has been suggested that these may have been the routes followed by some of the snails and crabs, and that some of the cryptozoic lines may also have utilized heavily vegetated areas as transition zones. Of these the nemertines are an example, well documented by recent studies. Other cryptozoic lines may also have moved landwards from mangrove swamps, and, although we know little about them as yet, further work in this area might be rewarding.

The evolution of terrestrial littorinacean gastropods and of terrestrial decapod crustaceans is thought from the fossil record to have occurred from Cretaceous to Tertiary times. For the soft-bodied groups, we have no reliable information about the timing of the terrestrial invasion from geological evidence. To be convinced that any of these groups may have moved on to land through saltmarshes or mangrove swamps, we must be sure that such environments existed at the appropriate time – a problem already mentioned for saltmarshes in Section 6.1. Looking at the fossil history of mangrove trees we find, in fact, that only the Combretaceae have been recorded as far back as the Upper Cretaceous (Chesters, Gnauck & Hughes, 1967; but see also Retallack & Dilcher, 1981). The other families did not appear until the Tertiary, and their spread probably took place between Upper Cretaceous and Miocene (Walsh, 1974). Use of mangrove swamps as routes to land would therefore have been just possible at the time postulated.

There is, however, a further possibility to be explored here. In Chapter 1, little attention was paid to the habitat of the various taxa of emergent plants that invaded the land, although it was noted that the 'Rhyniophyta' may have lived in fringing marine habitats. They may have provided an early equivalent of saltmarsh. It is likely that the majority of semi-aquatic plants evolved in fresh water; but it has been proposed that some of the larger gymnosperms forming swamp communities in the Upper Carboniferous were 'mangroves' – i.e. that they grew in salt water (Raymond & Phillips, 1983). Trees of the gymnosperm order Cordaitales alternate in the fossil record of coal-forests in Iowa (USA) with ferns and seed-ferns of freshwater origin. Although all these plants probably had aerial roots, possibly similar to the present-day *Rhizophora*, only the Cordaitales trees are thought to represent a community which characterized marine transgressions, while the ferns and seed-ferns appeared when fresh water dominated the area. It is not known how tall these particular species of Cordaitales were, but other species reached to heights of over 20 m. If this interpretation of some Upper Carboniferous forests as mangroves is correct, the possibility exists that many animal groups may have moved on to land by passing through the sheltering environment provided by them. Although we have no direct knowledge of the conditions within these early mangrove communities, it may be possible to extrapolate from what is known about present-day (angiosperm) mangrove swamps. This must be a high priority in any future research on the early origins of terrestrial invertebrates.

# 7

## Rocky and shingle shores as routes on to land

Their lives were a vivid illustration of how the dry barren land came first to be peopled with crawling animals. Where the limpets, anemones and sea urchins were still completely bound to the ocean, perishing miserably of heat and dryness when it left them for long, the Grapsoid crabs were able to forsake it for hours at a time. They were still creatures of the sea, but they were also land animals in process of transition.
Gilbert Klingel (1959) *Wonders of Inagua*, London: Robert Hale.

On coasts exposed to severe wave action, sandy shores alternate with hard rock surfaces and shingle barriers. Mobile sediments such as sand and shingle are mainly derived from offshore sources (Pethick, 1984), so that the type of shore in exposed situations is determined primarily by the availability of offshore sediment. Where this is lacking, rocky shores occur, and, since there may also be a lack of sediment supply in sheltered areas, rocky shores may sometimes be found even where there is little exposure to wave action. Overall, much of the variation in the physical character of rocky shores can be explained by variation in degree of wave action, as superbly described by Lewis (1964). Conditions in the intertidal zone also change markedly with height on the shore, and the vertical distribution patterns of organisms have given rise to wide-scale discussion about the importance of tidal levels in determining this distribution (e.g. Underwood, 1978). On a smaller scale, conditions on rocky shores also change with aspect (the direction in which they face), because this influences the desiccating power of the sun; with slope, which influences drainage rates; and with rock type, which determines the texture of the surface, and the types of refuges available in the forms of crevices, pits and overhangs. Once a community is established, there are also important biotic interactions such as competition and predation, each modified by the physical conditions of the shore. It can be seen, then, that rocky shores provide an enormous heterogeneity of habitats – much more so than muddy and sandy environments, and possibly more even than mangrove swamps. Before considering the animals that may have moved across rocky shores on to land, it is therefore essential to spend a brief time considering the environments involved.

## 7.1 The environment on rocky shores

The general physical characteristics of rocky shores have been discussed by many authors (e.g. Lewis, 1964; Newell, 1970). Alternating submersion and emersion caused by tidal action have understandably received prime consideration since they are responsible for the existence of an intertidal environment, and intertidal organisms live in vertical zones which can be related to tidal characteristics. Nevertheless, discussion about the primary cause of this zonation has occupied littoral biologists for decades. Two basic schools of thought on this subject still exist: one suggests that the physical factors associated with the tides control animal and plant distribution, while the other suggests that biological interactions form the most significant forces governing distribution. The first theory, championed for example by Lewis (1964), has led to the suggestion that species are limited on the shore by their capacity to withstand extremes of factors such as high and low temperatures and desiccation stress (Newell, 1970). The second theory, with supporters such as Stephenson & Stephenson (1972) has led to the contrary suggestion, that animals live at some particular level on the shore for various biological reasons determined by food sources, competition and predation, and that they have therefore become adapted to withstand appropriate changes in temperature and desiccation stress found at those levels (Underwood, 1979). An appreciation of these two opposing theories is essential for any understanding of modern intertidal biology, since they have strongly coloured the approaches of individuals to research on the causes underlying animal distribution, although the truth, as pointed out by Moore & Powell (1985) probably lies in a mixture of the two. A further twist is provided by the suggestion that most biological factors affecting distribution are themselves underlain by some form of physical control.

The intimate mixture of biological and physical factors controlling animal distribution on rocky shores is well shown by the interaction between mobile gastropod molluscs and macroalgae. Very few molluscs utilize macroalgae as a direct food source, although many eat the epiphytic microalgae growing on them, and take advantage of the different microclimate provided by the macroalgal fronds. In Britain, for example, *Littorina obtusata* lives mainly on *Ascophyllum nodosum*, eating the surface of the fronds and sheltering within the algal mass at low tide. The related *L. mariae* lives lower on the shore, on *Fucus serratus*, eating its epiphytic microalgae and again using the algal mass as a shelter against desiccation (Williams, 1989). Other gastropods such as *Lacuna vincta* settle preferentially on red seaweeds, which offer shelter and a food source at low levels on the shore (Fretter & Manly, 1977). The distribution of all these seaweeds is determined, in turn, by a mixture of physical and biological factors. For many of them the upper vertical limit on the shore is determined partly by their tolerance of aerial emersion during hot weather, but partly by competition with other algal species, and by their ability to survive the grazing pressure of invertebrates. The lower limits of the weeds are primarily controlled by relative growth rates (Norton, 1985). These factors underlying the distribution of macroalgae are therefore indirectly responsible for the distribution of the gastropods associated with the weeds, and the overall picture involves a very complex web of interactive factors.

Unfortunately, one of the major gaps in our knowledge of rocky shore ecology lies in the establishment of beach micro-climates. Although much has been written about the importance – and lack of importance – of physical factors such as temperature, desiccation and salinity stress, there appear to be very few measurements of actual climatic regimes on the beach. As will be discussed in Section 7.4, experiments on gastropods living high on the shore suggest that normal temperatures and humidities are unlikely to be important sources of mortality. The situation may be more complex than it seems, however, for as has been pointed out at length in Chapter 2, micro-climates can be very different from the general macro-climate, and the temperatures of *animals*, in particular, can be quite different from the temperatures as measured for the macro-climate. This will be further discussed in Section 7.7, in regard to the conditions experienced by high-shore isopod crustaceans.

Some indirect evidence about the relative harshness of rocky shore climates can be gathered by examining the difference between open-shore environments and those found in crevices, and by comparing the climate at the surface of shingle and boulders with that underneath the stones. These aspects are covered in the next two sections.

## 7.2     Crevice habitats

Irregularities in the rock surface offer protection against wave action and predation. Even small pits in a smooth rock face allow gastropods to avoid the full scouring force of waves, and empty barnacle cases provide another refuge, commonly used by isopods and gastropods.

Larger crevices in the rock itself offer shelter from waves and predators, protection from desiccation, and also the possibility of entering food webs that are somewhat isolated from those on the open rock face. Organic sediments collect within the crevices, providing food for detritivores. Filter-feeding animals within the crevices trap suspended sediments, and then themselves provide a food source for small predators within the crevice (Glynn-Williams & Hobart, 1952).

Physical conditions within the crevices can be very different from those on the open rock face. In a study on the south coast of Britain, Morton (1954) showed that crevices in slate had a variety of temperature and humidity regimes. Shaded crevices usually had temperatures lower than those of the open rock face, but crevices facing the sun were hotter than the outside. At about mean high water neaps, shaded crevices remained near sea temperature, at about 12–13 °C, while air shade-temperatures in the open rose to 20 °C, and sunny crevices at the same height reached 23 °C. In the shaded crevices the relative humidity remained between 95% and 100%, while in sunny crevices it fell to about 80%.

The shaded crevices examined by Morton contained a wide diversity of fauna, from both marine and terrestrial origins (Fig. 7.1). Isopod crustaceans were abundant, especially *Ligia oceanica*, which retires to humid micro-climates in the day but emerges to forage at night (see Section 7.7). The crustaceans were accompanied by pulmonate gastropods such as *Leucophytia* and *Ovatella* (see Section 7.8). Both these groups may also be abundant on shingle and boulder shores, and their presence emphasizes the similarity of conditions between the interstices of shingle and crevices in solid rock.

The distribution of animals in the crevices is related to many factors, but the most important are probably relative humidity, and associated temperatures, and food supply. At high tidal levels, crevices are not sheltered by dense algal growths, so they tend to dry out, and there is relatively little food. In contrast, crevices lower on the beach may, in the north temperate zone at least, be protected by fucoid seaweeds which also provide a rich source of organic detritus. This detritus also clogs up the crevices, thereby altering the physical properties of the niches. As a result of these interacting factors, some animals such as the pulmonate *Otina*, the centipede *Scolioplanes* and the mites, *Bdella* spp. are restricted to high-level crevices, whereas low-level crevices contain a variety of soft-bodied animals such as polychaetes and nemertines. In crevices at around mean high water neaps, in intermediate conditions, there is a wider diversity of species, where the two faunas overlap in distribution. To some extent the

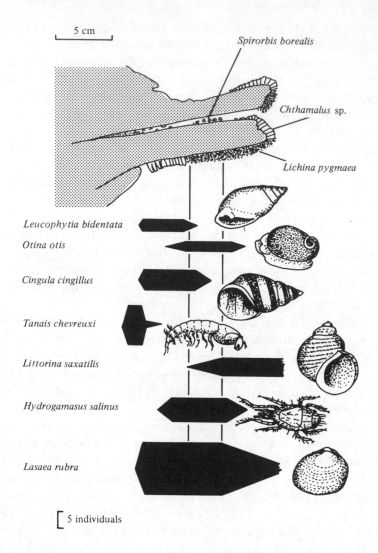

Fig. 7.1. Distribution of some of the fauna in an intertidal rock crevice in slate, at Wembury on the south coast of England. The vertical lines show the zone of *Spirorbis borealis*. Most of the species shown are of marine origin, but all are probably able to breathe air or to close up and metabolize anaerobically during low tide. After Morton (1954).

high-level fauna can be termed a 'terrestrial' component (Glynne-Williams & Hobart, 1952), since the majority of its species breathe air, and have been directly derived from fully terrestrial ancestors. It contains, besides those species listed above, collembolans, dipterans, beetles and pseudoscorpions. Some animals of undoubted terrestrial origin can, however, colonize a more extreme vertical range: hemipterans such as *Aepophilus* spp. are found down to the level of mean low water springs. Likewise, the low-level fauna can to some extent be equated with an 'aquatic' component, since most of the species involved probably breathe dissolved oxygen and have direct origins from marine ancestors. Here are included some of the brittlestars, tanaids and bivalve molluscs, but again there are *some* species that show a wider distribution. The bivalve *Lasaea*, for instance, is found up to mean high water neaps, as is the polychaete *Spirorbis borealis* and the tanaid *Tanais chevreuxi*.

In summary, crevice faunas provide a striking demonstration of the abilities of marine animals to adapt to aerial conditions, and of terrestrial animals to adapt to submergence. In turn we may expect that they will provide some interesting angles on the ways in which such adaptations have evolved.

## 7.3    Shingle and boulder-shore habitats

In Chapters 5 and 6, we considered environments provided by mobile sediments in the form of muds and sands. To some extent the conditions intermediate between these and rocky shores are provided by shingle ridges and boulder shores. Here the individual rocks have varying degrees of mobility depending upon their size and the degree of wave action to which the shore is exposed. In general, shingle bars have quite well-sorted size fractions, and the shingle is very mobile so that the fauna is restricted. Boulder shores, on the other hand, have a wide range of sediment sizes, from sand and mud particles up to 'boulders' which on the Wentworth scale of size are classified as greater than 256 mm in diameter. The extreme heterogeneity of particles on boulder shores means that many types of niche are probaby available, but surprisingly few studies have concentrated on the characteristics of the range of micro-habitats.

On a boulder shore in Scotland, Agnew & Taylor (1986) provided the first comprehensive account of the variations in physical and chemical conditions over the year (Fig. 7.2). They found that under large cobbles, 200 mm in diameter, fluctuations in temperature and humidity were considerably reduced in comparison with those in the open, especially in summer. Maximum variations in temperature, in a 24 h period, even on the

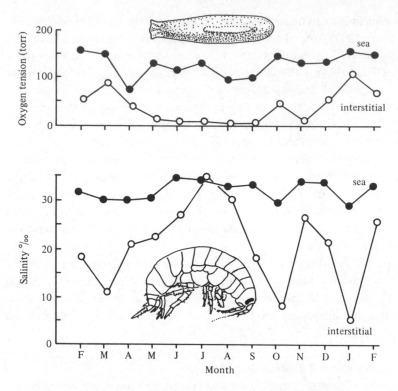

Fig. 7.2. Seasonal variations in salinity and oxygen tension on a boulder beach in Scotland, at about mean high water neaps. Filled circles show sea water. Open circles show mean values in interstitial water. Despite large changes in both variables, the flatworm *Procerodes littoralis* and the amphipod *Chaetogammarus pirloti* were abundant. After Agnew & Taylor (1986).

upper shore (at the level of mean high water neap tides) were only 10 °C under the cobbles, but 20 °C in the open. Variations in humidity were limited to 30% relative humidity under the cobbles, with a minimum of 70% R.H. On the open beach, in contrast, variations over 24 h were as much as 45%, and values sometimes fell as low as 50% R.H. At high levels, above mean high water neaps, humidity even under stones fell to 50%, so that a much harsher micro-climate prevailed. As with other types of shore, there is probably a very harsh region at the top of shingle and boulder beaches where physical fluctuations are more extreme than either those lower down or those on land. This relates especially to the lack of vegetation at the top of the shore.

Agnew & Taylor (1986) also measured fluctuations in salinity, oxygen tension and pH. The interstitial water beneath boulders showed wide changes in salinity over the year. In autumn and winter it decreased to as little as $5°/_{oo}$, whereas in summer it stayed relatively constant at around $30°/_{oo}$. The decrease was due to run-off from the land, because the salinity of the tidal water stayed fairly constant at about $30°/_{oo}$ throughout the year. The oxygen tension of water under the stones varied from saturation values to total anoxia. This variation was mainly due to varying degrees of oxygen absorption by belts of decaying seaweed (wrack) deposited on the beach. The wrack also probably forms a significant part of the food source of the inhabitants of the boulder shore, just as it does on intertidal sands (see 5.2.5), so it is apparent that it provides a dominant influence. Its distribution may also influence pH, although seasonal changes in pH were rather erratic.

The habitats under cobbles could be divided into those with standing water and those without. The wet habitats resembled rock pools in their changing physical and chemical variables (see 2.3). The dry habitats were very similar in character to the crevices just discussed. Presumably for this reason, stable shingle and boulder shores have a fauna which shows much in common with that of rocky shore crevices. In particular, isopod crustaceans such as *Ligia* spp. are common (see 7.7), and there is often a contingent of pulmonate gastropods (see 7.8) and nemertines (7.9).

## 7.4     Rocky shores: the adaptations of littorinid gastropods
One of the most universally distributed animal groups on rocky shores is the gastropod family Littorinidae. The littorinids are usually accompanied by limpets, both prosobranch and pulmonate, by trochids, and by neritids. All of these are primarily grazers, feeding on microalgae and sporelings on the rock surface. On many shores it is the presence of these grazers that reduces the cover by macro-algae.

In the north temperate zone, on European shores and the north-east coast of north America, *Littorina littorea* and *L. obtusata* are common in the mid-littoral. Higher up the shore, the various species of the *Littorina saxatilis* complex are found, together with *Melarhaphe neritoides*. These species have received much attention from European researchers, and show a variety of adaptations to life on the shore. In the tropics and subtropics, a wide variety of littorinids is found: *Tectarius muricatus* high up on shores in the Caribbean, for instance, and *Littorina aspera* high up on the Pacific coast of Panama. In the south temperate zone littorinids are again widespread, with *L. unifasciata* and *L. acutispira* and many others on the east coast of Australia.

Littorinids are found at all levels on the shore, but the major interest here is concerned with those species living at high levels: How are they adapted to withstand the conditions there? Are they limited in their vertical distribution by the harshness of climatic factors? Are they well adapted to the physical conditions of specific regions because they need to live there in order to feed and reproduce? Has the rocky shore environment been for littorinids a pathway on to land? Some of these questions can be approached by considering in turn some north temperate, south temperate and tropical species of littorinid.

On British shores, the patterns of vertical distribution shown by the eight species of common littorinid have been well documented (Fretter & Graham, 1980). Ignoring the species adapted for life on macroalgae, the commonest species are *L. littorea* (mid-shore up to mean high water neaps), *L. saxatilis* (up to mean high water springs) and *M. neritoides* (up to extreme high water springs and above). The resistance of these littorinids to desiccation increases in those species higher on the shore, but this is far from providing a simple story of steadily increasing adaptation to emersion with vertical height; and the discovery that *L. saxatilis* is only part of a complex of species with a variety of characteristics has not clarified the situation.

The reproductive adaptations of the group demonstrate the plasticity of littorinids, and the variety of ways in which species can be adapted to cope with environmental stress. *L. littorea*, the large mid-shore species, produces planktonic egg capsules from which hatch planktonic veliger larvae. *M. neritoides*, a small high-shore species, shows the *same* reproductive strategy, and size for size produces the same number of eggs. Yet *L. saxatilis*, intermediate in size and vertical position, is ovoviviparous, brooding its young. The picture is further complicated by some species closely related to *L. saxatilis*. *L. nigrolineata*, on the mid shore, lays benthic egg capsules from which hatch young snails. Exactly the same strategy is shown by *L. arcana*, usually found between mean high water neaps and mean high water springs. Finally, *L. neglecta*, a very small species usually found in dead barnacle cases, broods its young like *L. saxatilis*. These modes of reproduction seem completely unrelated to vertical position on the shore, and the factors governing reproductive mode may be related more to adult size and perhaps to varying necessity for dispersal (Underwood, 1979).

A clearer picture is obtained by taking a broader view and examining littorinids worldwide. On Australian coasts, for instance, there are four common species of littorinid: *Nodilittorina pyramidalis* at the top of the shore, then *Littorina unifasciata* and *L. acutispira*, and *Bembicium nanum* lower down. Of these, the first three all produce planktonic egg capsules, while *B. nanum* lays benthic eggs from which hatch pelagic larvae. This picture is more common worldwide than the European one, and ovoviviparity is seen overall to be of rare occurrence: it occurs only in *L. saxatilis*, *L. neglecta*, a lagoonal species *L. tenebrosa*, and the mangrove littorinids (Underwood & McFadyen, 1983). Almost all high-shore littorinids, in particular, produce planktonic eggs which hatch into planktonic veliger larvae. It is possible that they need to do this because they live in an unpredictable environment of physical extremes, where populations can easily be annihilated. High-shore snails also tend to be relatively restricted in their movements, and a planktonic dispersal phase may be essential on this count. This dependence upon planktonic eggs when living at the top of the shore suggests that such a line would be unlikely to give rise directly to a terrestrial line, which must depend upon the laying of eggs in soil, or upon brooding. The distribution of ovoviviparous forms on marine shores does not, in contrast, explain the lack of ovoviviparity in terrestrial littorinaceans: although on rocky shores these forms are limited to the mid-shore, the mangrove species reach high levels. It is perhaps surprising that these forms never produced terrestrial ovoviviparous lines, especially as this mode of reproduction seems to be the norm in terrestrial isopod and amphipod crustaceans.

Returning to Australian rocky shores, *Littorina acutispira* has been studied by Underwood & McFadyen (1983). This is a very small snail, less than 3 mm in height, which lives on open rock and in pools at high levels. It can withstand extremely high temperatures and low humidities, as is necessary on open rock faces: experimentally, it showed little mortality even after 24 h at 37.5 °C and 20% R.H. Since these are conditions to which it would never be exposed in nature, it was suggested that temperature and desiccation are unlikely to be major causes of mortality on the shore.

This subject has been examined in more detail by Garrity (1984) for gastropods on the tropical Pacific coast of Panama. Here *Littorina aspera* lives high on the shore, in the splash zone. On clear sunny days, the rock in this area can reach temperatures over 45 °C. Snails are active only when the rocks are damp, and aggregate in crevices and on vertical cliffs when the rocks are dry. While inactive, the tissues of the snails may be cooler than the adjacent rock face by as much as 5 °C, presumably due to evaporative cooling. The snails retract into the shell, seal the operculum, and attach the lip of the shell to the rock with mucus. Thus orientated, the spire points upwards, so that the shells receive the minimum possible amount of solar radiation, and also have the minimum area in contact with the substrate.

The importance of these behavioural procedures in reducing or limiting body temperatures was shown experimentally by moving the snails. Individuals moved from crevices to open surfaces, or those whose orientation was altered, all increased in temperature by about 4 °C. In other species, such as *L. modesta*, these manipulations significantly increased mortality, but this did not occur in *L. aspera*. Experimental removal of the operculum in *L. aspera* did produce an enormous increase in mortality, attributable to an increased loss of water.

These experiments suggest that, at least in the tropics, where the shore is bare and has no protective canopy of macroalgae, littorinids may be living very near their lethal temperatures. This does not mean that high temperatures and desiccation normally contribute to mortality. Rather, the snails are exceptionally well adapted to regions of extreme physical stress, and can well withstand the rigours of the tropical intertidal regime. Nevertheless, there is doubtless a physiological barrier to the movement of gastropods above the splash zone on these tropical rocky shores. Not only will the temperatures there be even more extreme, but the supply of water becomes even more irregular. As suggested for several previous shore types, there is here a region in which physical stresses are greater than those on land or in the lower intertidal zone. This barrier is so extreme in the tropics that it seems unlikely for animals to have moved directly across it.

Further examples of extreme ability to withstand rigorous conditions can be found in the subtropics. *Tectarius muricatus*, for instance, found at very high shore levels in the Caribbean, has been found to survive for up to 17 months dry in the laboratory, and then to respond rapidly to water. Such abilities are also characteristic of some of the terrestrial Pomatiasidae, especially those living in limestone areas, suggesting a possible direct relationship. Any suggestion of direct origin of terrestrial forms from rocky shore ancestors is, however, countered by the dissimilarity of reproductive mechanisms mentioned above: *Tectarius* produces pelagic larvae, and is inescapably linked to the sea.

## 7.5    The importance of crevices: *Petrobius* and other insects

In Section 5.2.7, the possible origins of terrestrial insects from interstitial ancestors was discussed. In some sense, the crevice habitat can be considered an extension of the interstitial habitat, with rather larger pore spaces, so that it is appropriate to examine the life-styles of present-day primitive insects that live in marine and maritime crevices. The best-known examples are some of the apterygotes, specifically the machilid Thysanura.

The majority of machilids are found inland, in cryptozoic habitats,

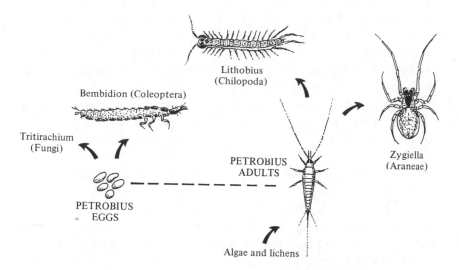

Fig. 7.3. Part of the food web involving the marine high-shore thysanuran, *Petrobius brevistylis*. Although this species is almost confined to the maritime zone, and feeds on maritime plants, its major predators are of terrestrial origin. From observations by Delany (1959).

especially under rocks and often in forested areas. Some species have adapted to the region near the top of marine shores, and these give a good indication of the adaptability of apterygotes to a region that, as we have seen, often forms a severe barrier to animals moving from the sea to land. In Britain, two species are found in this high-shore area: *Petrobius maritimus* and *P. brevistylis* (Davies & Richardson, 1970). Both feed by scratching algae, lichens and mosses from open rock surfaces at night, but retire to crevices during the day. *P. brevistylis* feeds on large, continuous expanses of solid rock, and is therefore usually found in the daytime in narrow crevices in cliff faces, with a width of 5–10 mm. *P. maritimus* feeds on smaller areas of rock such as boulders, and is more often found in shingle banks and under loose stones. Both species require dry resting places, and both also need some degree of elevated salt supply: their commonest distribution is near high water mark, and, when found inland, the sites are in south-west Britain, where prevailing winds blow salt off the sea. Although both are thus restricted to maritime regions, their trophic interactions link them much more with terrestrial food webs (Fig. 7.3): both adults and eggs are subject to predation from terrestrial sources.

The structural and physiological adaptations of thysanurans have been discussed in Section 5.2.7, in relation to the interstitial habitat. Many of these adaptations can be seen, in the case of *Petrobius*, to be equally adaptive for life in crevices. In particular, the use of abdominal sacs for rapidly taking up water in an environment where water is both scarce and appears only spasmodically must be advantageous. The ability of ma-chilids to live at the top of marine shores demonstrates the adaptability of some of the apterygote lines. Invasion of terrestrial crevices and cryptozoic niches from maritime equivalents by insect ancestors may thus be judged of equal importance to their invasion via the interstitial route.

### 7.6    Shingle and boulder habitats: the crabs *Cyclograpsus* and *Pachygrapsus*

The life-style of the high-shore thysanuran insects can instruc-tively be compared with that of high-shore decapod crustaceans. Two examples will be used here, one from New Zeland, and one from the west coast of north America.

In New Zealand, the upper levels of shingle beaches are occupied by the grapsid crab *Cyclograpsus lavauxi*. While the *Petrobius* spp. just discussed live *above* high water mark, *C. lavauxi* lives just *below* it. It is mainly nocturnal, and retires during the day to cavities beneath boulders and shingle. These cavities are not as dry as those occupied by *Petrobius*, and they are usually floored by sediments of small particle size (Bacon, 1971). In general, larger individuals are found higher on the shore than juveniles, but in the summer months the whole population tends to move downshore (Pellegrino, 1984), possibly indicating an avoidance of desiccation stress. *Cyclograpsus* is sometimes active in air, but is thought to forage mainly under water, thus differing from *Petrobius*, which avoids water completely.

The adaptations of *Cyclograpsus* to its high-level habitat have been investigated by Innes *et al.* (1986). The crabs are often not covered by the tide for several days at a time, and may be found with their gill chambers filled with air so that they cannot re-circulate branchial water like many other grapsids. Their resting rate of oxygen consumption in air is very similar to that in water, but when forced into activity, maximum rates of oxygen consumption are much greater in water, showing that the uptake mechanisms are primarily adapted for use in water (Fig. 7.4). This is normal for crabs, because the gill leaflets have insufficient support to allow them to stay separated in air, and the surface area available for gas exchange declines greatly when the leaflets collapse together.

*Cyclograpsus* can tolerate a loss of over 30% of its body water, which is high for a decapod. It is therefore able to spend a large proportion of the tidal cycle resting in its high-tidal habitat, breathing air, and waiting for

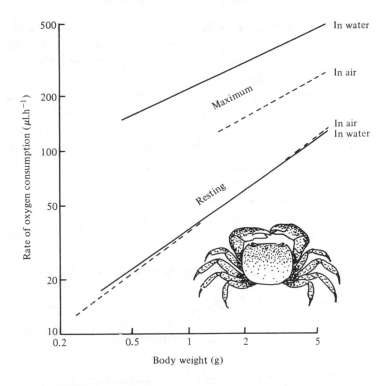

Fig. 7.4. Oxygen consumption by the New Zealand crab *Cyclograpsus lavauxi*. Rates when resting are very similar in air and water, but when the animals are stressed, much higher rates are reached in water. After Innes *et al.* (1986).

appropriate nocturnal high tides before it becomes active. It can then forage over the beach, and presumably exploits a niche that can be tolerated by few other marine animals.

Many grapsids have become adapted to yet more terrestrial habitats than *Cyclograpsus*, but this species gives us an indication that intermediate environments in the invasion of land may well have involved large crevices. These, especially in shingle, maintain humid micro-climates with large voids, and permit the upshore movement of large animals that could not survive permanently in the open. For decapods, which have a relatively permeable integument, this line of invasion could have been an important one.

This point is reinforced by a consideration of the American west-coast crab *Pachygrapsus crassipes*. This species is found from low water up to extreme high water, but like *Cyclograpsus* it hides under rocks and in large crevices during the day. When it does emerge during the day, it stays close to refuges. It usually comes out at dusk and remains out through the night, but in contrast to *Cyclograpsus* it feeds at low tide, out of water (Hiatt, 1948). *Pachygrapsus* is primarily a herbivore, feeding on the algal mats provided by *Ulva* and filamentous species at the edge of tide pools, at all levels on the shore, including the supralittoral. It is also a scavenger, and on occasions acts as a predator. It can run fast, and pursues animals such as sea slaters (*Ligia* spp.), and is rapid enough in its responses to catch flies. *Pachygrapsus* differs from many crabs in that it has an acute awareness of the environment. It uses visual cues rather than chemical ones, as shown by its reactions to people: it retreats rapidly to a refuge in response to a moving person 25 m away. It short eyestalks are elevated out of protective grooves when scanning the shore, in continual surveillance. This contrasts strongly with crabs active underwater, which use chemical cues for their major sensory input.

In summary, *Cyclograpsus* provides a good model for the adaptation of crabs to the high-shore habitat but this is very much a passive reaction to life in air; the crabs tolerate the aerial phase, but wait for the returning tide to commence activity. *Pachygrapsus* demonstrates more active adaptation to the higher part of the shore: it is active out of water, has a varied food source, and is a very active runner. It may provide a model for the ways in which some crab families have moved on to land.

### 7.7     Shingle and boulder habitats: the isopod *Ligia*

The origins of terrestrial isopods have been discussed in Section 4.7.1, in which a route directly from sea to land was postulated. In Section 5.2.8, the ecology of the sand-dwelling Tylidae was considered. More widespread, however, is the family Ligiidae, which contains species that are common on rocky shores. They often occupy habitats similar to those of the insect *Petrobius* spp. and the crab *Cyclograpsus*, in crevices and under boulders and shingle at high tidal levels. Once again, the micro-climates here provide refuges to which the isopods can retire in the day, and from which they emerge at night to forage.

The work of Edney (1953) on *Ligia oceanica* is of particular interest in relation to the nature of the micro-climates found in shingle and in rock crevices. In general, *Ligia* is thought to be active at night, like many of the crevice animals just discussed. On hot days, however, Edney observed *Ligia* walking on the *surface* of the rocks, having emerged from its normal refuges below. When temperature measurements were made in various niches, and compared with those of both living and dead animals, the reasons for this daytime emergence became clear. Normally, the spaces

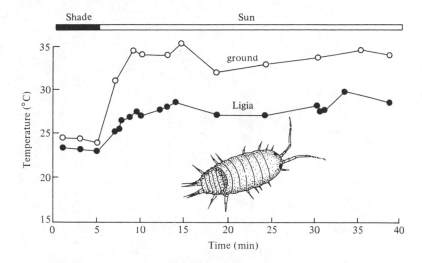

Fig. 7.5. The internal temperature of living *Ligia oceanica* and the ground, in this case a wooden surface. Relative humidity was 45–52% and wind speed 200 cm.s$^{-1}$, so that evaporative water loss cooled the isopod by as much as 7°C. After Edney (1953).

beneath shingle stay cool, but on sunny days the air there heated up to 30°C, which is very near to the lethal temperature for *Ligia*. Since this air was nearly saturated, no evaporation from the bodies of the *Ligia* could occur, so that they would inevitably have heated up to 30°C. Instead, they emerged on to the open rock, which was itself at a temperature of 34–38°C. Here the relative humidity was only 70%, and animals lost water by evaporation and convection. Their body temperature was thereby reduced to about 26°C, well below the lethal value (Fig. 7.5). In this particular case, then, although physical conditions in the crevices appeared to be more suitable than those outside, the animals survived better by leaving the crevices.

In the long term, *Ligia* must return to the crevice, or find a water supply, because evaporative and convective cooling will soon lead to an unacceptable degree of desiccation. For rocky shore isopods, which have little physiological or structural defence against water loss, the crevice refuge, or the adoption of a burrowing habit, were essential intermediates in the movement from the intertidal zone on to land.

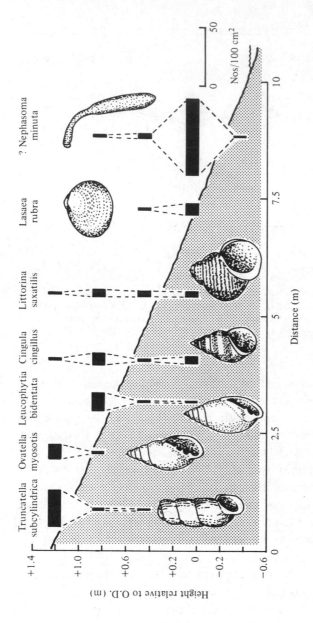

Fig. 7.6. Distribution of some molluscs (and one sipunculid) in the shingle beach of a brackish-water lagoon (The Fleet) in southern England. High-tidal species such as *Truncatella* and *Ovatella* are rarely submerged, and live in the humid, detritus-rich spaces between the pebbles. Others are restricted to lower zones, where they are accompanied by a more diverse fauna including the sipunculid *Nephasoma*. This section is taken through an area where oxygen-rich water emerges from springs at about the level of ordnance datum. Several of the species shown are also present in crevices on rocky shores (Fig. 7.1). After Little *et al.* (1989).

## 7.8 Shingle and boulder habitats: pulmonate gastropods

Storm beaches, consisting of small shingle and pebbles, are generally too mobile to allow many animals to live in them. Restricted areas of these beaches, on the landward side, may become stable, and accumulate sufficient fine sediments and organic detritus to build up communities of detritivores. One example has been shown in Britain in the Chesil Beach. This storm beach forms the outer border of a lagoon, The Fleet. At several places on the shores of this lagoon, water seeps continually out of the storm beach, and in these seepage areas a unique assemblage of pulmonate and prosobranch gastropods is found (see Little *et al.*, 1989). High on the beach, above the seepage line, are the pulmonates *Leucophytia* and *Ovatella* – well-known species in shingle habitats, adapted for breathing air, and only occasionally covered by the tides. With them is a variety of prosobranchs, including common rissoids such as *Cingula cingillus*, but also rare species such as *Truncatella subcylindrica* and *Paludinella littorina* (Fig. 7.6). This high diversity demonstrates that with an abundant supply of detritus (derived in this case from *Zostera*), the shingle environment can be attractive to several gastropod lines. The dependence upon decaying weed is reminiscent of the food webs found in the interstitial sand habitats discussed in Chapter 5.

Another sheltered-shore boulder beach has been discussed by Morton & Miller (1973). Near Auckland, in the North Island of New Zealand, is a volcanic island called Rangitoto. Here the scoria boulders provide a refuge which is often covered by decaying algal wrack. In the spaces below the boulders is an assemblage of pulmonates. One, *Rangitotoa*, was named for the locality, but others such as *Leuconopsis* are more widespread, and are often found with prosobranchs. Most of these species are typical cryptozoic animals, confined to high humidities and dark regions: exposure to dry air and increased temperature is rapidly lethal.

In previous sections (4.8.1, 6.1.2), the possible invasion of the land by marine pulmonates has been discussed with reference to saltmarshes and mangrove swamps. Crevice habitats provide another intriguing possibility for a route to land, in which pulmonate gastropods could have moved from a marine cryptozoic niche directly to a terrestrial equivalent.

## 7.9 Shingle and crevice habitats: the nemertine *Acteonemertes*

Nemertines are the epitome of cryptozoic life, requiring high humidity, a reliable water supply and constant temperature. Their invasion of the land therefore seems remarkable, but terrestrial nemertines have evolved on several occasions. Apart from the probable origin of the *Geonemertes* group in saltmarshes and mangroves (discussed in Section 6.3.5), there is good evidence that *Acteonemertes bathamae* has moved on to land *via* shingle beaches. This species is found at the top of shingle shores in New Zealand, in the company of the crab *Cyclograpsus lavauxi*. It has now also been recorded on land, at a height of 30 m above the sea (Moore, 1973), and can evidently withstand terrestrial conditions. There seems little doubt that this species has invaded terrestrial cryptozoic habitats directly.

## 7.10    Conclusions

As with sandy shores, rocky and shingle shores appear to present a relatively inhospitable front to invading animal lines. Very few groups have been successful high on rocky shores. Of these perhaps the commonest are the gastropod molluscs and the cirripede crustaceans, the barnacles. The latter, however, are of little direct interest in the present context because as filter feeders they cannot survive in the terrestrial habitat, while the prosobranch gastropods, although nicely adapted to the rigours of high-shore existence, have probably not been able to cross the harsh barrier between shore and land. Instead, the animals more likely to have moved landwards are some of the more insignificant lines that have made use of crevice environments. These crevices have provided conditions in which cryptozoic animals can avoid the problems of desiccation, as well as those of predation, which characterize the open shore. Using these sheltered conditions, it may have been possible for several cryptozoic lines to avoid the important barrier region at the junction of the intertidal zone and the terrestrial zone.

In this movement across rocky shores, one of the major problems to consider is the supply of food, since at the top of rocky shores there is usually little growth of either marine or terrestrial plants. Here the similarities with sandy shores emerge again, as most food webs at the top of rocky shores are fuelled by deposits of detached marine algae – the wrack beds. This source of detritus, mostly consumed at night, when crevice-dwelling animals can emerge from their cryptozoic hideouts, has probably provided the 'incentive' for animals to invade an otherwise relatively hostile environment. At the same time, it has made possible the next step, on to land.

# 8

## The freshwater route to land

When I looked across the pond . . . all the earth beyond the
pond appeared like a thin crust insulated and floated even by
this small sheet of intervening water, and I was reminded that
this on which I dwelt was but *dry land*.
H. D. Thoreau (1854) *Walden, or Life in the Woods*.

In Chapters 5 to 7, discussion of the environments in which
terrestrial groups may have evolved has centred on the various types of
marine intertidal zone. It has been suggested from this discussion that a
large proportion of terrestrial invertebrate lines evolved directly from
marine or estuarine ancestors, including those ancestors adapted to
saltmarshes and mangrove swamps. These 'direct' routes on to the land, as
discussed in Chapter 4, often led to the retention of 'tolerance' mechanisms
by the terrestrial lineages. Estuarine and other brackish-water habitats,
however, have a further significance for the evolution of terrestrial lines
because they have provided the route for the invasion of fresh waters. In
Chapter 4 it was suggested that freshwater lines invading land mostly
retained 'regulatory' mechanisms, and hence differed radically from direct
invaders from the sea.

The present chapter deals with this 'indirect' route to land, via fresh
water. To begin with, it introduces the problems of life in brackish waters,
compared with those found in the sea. It then considers conditions in
various types of fresh water, and takes examples from a number of animal
groups thought to have moved from fresh water on to land.

### 8.1    Brackish-water environments

There are two primary areas in which the sea interfaces with fresh
water, providing brackish-water environments. These are estuaries and
coastal lagoons – each with their own characteristics, but with no hard and
fast dividing line between them. They can be regarded as two separate
routes to fresh water, and are now considered in turn.

### 8.1.1   Estuaries

Estuaries provide a wide variety of conditions, and as emphasized by Barnes (1984) they form a continuum from areas where small rivers discharge into a restricted volume of sea water to areas where large rivers flow directly into the open sea. In terms of salinity, this means that the major characteristic of estuaries is variation: salinity may change with distance up the estuary, with depth, with season, and with each high and low tide. As in the marine littoral zone, the major response of estuarine animals has been to adopt mechanisms of tolerance. It is true that a wide variety of crustaceans show osmoregulatory ability (e.g. Mantel & Farmer, 1983), but in general the changes in salt concentration are so large that it is energetically very expensive to attempt to osmoregulate (Potts, 1954). The major exception here is shown by the migratory vertebrates, both those that live in the sea and breed in fresh water, and those that live in fresh water but migrate to the sea to breed. Both of these have extraordinary abilities of salt and water balance (Evans, 1979).

The variation in salinity within estuaries has often been regarded as the major determinant of faunal composition (Remane & Schlieper, 1971). The salinity regimes are certainly stressful, and many groups of animals are unable to withstand such conditions. Several other factors are, however, very important in controlling animal distribution (Barnes, 1984; McLusky, 1981). In particular, the substrate types must be considered, and these are intimately linked with water movements. Tidal currents in estuaries may be very high, and in the fast-flowing stretches the substrate will consist of sand, or coarser particles. Since currents reverse on each ebb and flood tide, these deposits may be very mobile, providing little scope for other than opportunistic animals capable of living without any fixed home territory. In areas where these currents are slow, fine particles will be deposited, and more permanent burrows and territories may be established. The types of animals described in Chapters 5 and 6 will then be established in mudflats, saltmarshes and mangrove swamps respectively.

Overall, the estuarine environment has always been regarded as a stressful one. In some senses we can envisage it as a 'filter', allowing only the most adaptable lines through it into the freshwater environment. Some groups, such as the echinoderms and the cephalopod molluscs, have never produced freshwater representatives. Others, such as the vertebrates, the bivalve and gastropod molluscs, and several crustacean groups, have diversified in estuaries and have become extremely successful in fresh water. As will be seen later, some of these groups have utilized their freshwater adaptations to invade the land.

### 8.1.2    Coastal lagoons

The conditions within coastal lagoons have been excellently described by Barnes (1980). In terms of salinity, at least, most lagoons are more stable than estuaries, because the exchange of water with the sea is restricted by the relatively small size of their entrances. Salinities may still fluctuate widely over the year, but not so much on a diurnal or tidal scale. There are exceptions to this rule, particularly if the lagoon contains a salt-water reservoir below a halocline (e.g. Barnes *et al.*, 1979). In this case, salinity can change rapidly over short time-periods, due to the mixing of surface and bottom water by the action of the wind. This property has not, however, been shown to be widespread.

Most lagoons show little regular change in water level, again because of the restricted interaction with the sea. There is therefore only a limited intertidal zone, and salt-tolerant fringing vegetation such as mangroves, saltmarsh and reedswamp may merge directly with terrestrial communities. Here, then, there may be a direct opportunity for animal lines to move from lagoons to land via vegetated zones, as described in Chapter 6.

The stability of lagoonal conditions may also have provided an easier route by which animal lines could invade fresh water than any route through estuaries: most lagoons have a freshwater inflow, and preliminary adaptation to lagoons could have allowed animals to move into freshwater habitats. Subsequent movements from fresh water to land are now considered in detail.

### 8.2    Freshwater environments

Since all fresh water originates from some form of precipitation such as rain or snow, it might be supposed that its salt content would always be extremely low. In practice, all precipitation has a substantial salt content, highest near the sea and lowest inland, suggesting that the salt is derived mainly from airborne particles of sea salt (Hutchinson, 1975). The pH of rain is normally below 7, due to solution of carbon dioxide, and its acidic nature tends to dissolve soils and rocks, especially those with high carbonate contents such as limestone. Although, therefore, the principal ions in rain are usually sodium and chloride, the most abundant component ions of fresh water are often calcium and carbonate.

For animal lines invading fresh waters from brackish waters, the two major chemical changes experienced are consequently: first, a decrease in salinity from a figure that may be as high as $35^\circ/_{oo}$ (that of sea water) to values of less than $1^\circ/_{oo}$; and secondly, a change from a sodium and chloride dominated composition to one dominated by calcium and carbonate. Coupled with these two drastic changes is a tendency for the composition to be much more stable than in estuaries or coastal lagoons: once physiological adjustment to freshwater chemistry has been made, problems of life there are much less severe than in brackish water.

As well as differences in chemistry, freshwater environments show many physical and biological differences from brackish waters. Freshwater environments and their biological communities are very diverse (e.g. Gray, 1988), but here we will consider briefly only the two major subdivisions, running water and still water.

### 8.2.1     Rivers and streams

At the point where rivers flow into, or become estuaries, their chemical content is usually far removed from that of rain: their water contains, besides calcium and carbonate derived from solution of bedrock, relatively high levels of nutrients such as phosphate, nitrate and silicate, and both dissolved and particulate organic matter. Much of the organic matter is derived from leaf-fall of trees overhanging the streams in their upper reaches, and it provides a major source of food for invertebrates. Such a rich food source can only have been available after the evolution of coniferous forests in Carboniferous times, and it is possible that the quality of the supply vastly changed as the angiosperms evolved in the Cretaceous. For invertebrates invading fresh water before the Cretaceous, the majority may have had to depend for food upon the primary production of algae within the water itself.

With distance upstream, the total dissolved salts decrease in concentration. Accompanying this decrease are several physical changes, the major one being a change in the relationship between stream flow and water depth. As emphasized by Townsend (1980), the absolute water velocity usually decreases towards the source, but, because the water depth decreases, the shear stress on the bed of the stream is highest in the headwaters. For benthic animals, therefore, major adaptations are concerned with clinging to the substrate, and streamlining water flow over the body surface. The substrate itself contains few small particles, again because of the high shear stress.

In all sections of rivers there is a mosaic pattern of different habitats, equivalent to those described for many marine shores. Particularly in smaller rivers, for example, shallow areas of fast flow, known as riffles, often alternate with deeper areas of slower flow, the pools. Riffles and pools have quite different faunas, adapted to the physical conditions present. The substrate in the riffles usually consists of gravels and larger stones, and in this complex matrix there is a large and diverse fauna. The presence of silt reduces the fauna, and in the pools the fauna of fine sediments is much less diverse (Hynes, 1970).

In smaller rivers and streams, too, benthic macrophytes may be abundant – a marked distinction from estuaries, many of which have few submerged macrophytes because their water is very turbid. Diversity of emergent macrophytes is also much greater at the edges of freshwater streams than in estuaries, where saltmarshes contain relatively few species.

Overall, the diversity of habitats in rivers is very much greater than in estuaries. Coupled with the more stable chemical conditions, it is perhaps to be expected that animal lines, having once invaded fresh water, should diversify there more than in estuaries. In fact, the number of invertebrate species in fresh water is very much greater than in estuaries, but the number at the present day is composed to a large extent of insects, particularly larval stages. These have probably invaded fresh water from a terrestrial base, and not via estuaries. Apart from the insects, the major freshwater groups are Crustacea, Annelida, Platyhelminthes, Mollusca and Vertebrata. The species representing these groups in rivers and streams are all adapted to conditions in which water is continually replaced from upstream, i.e. to conditions of high oxygen concentration. Most have gills to obtain oxygen, or have very high surface area: volume ratios which allow them to obtain oxygen by diffusion through the body surface. In either case, movement on to land would produce immediate respiratory problems, because neither mechanism would be satisfactory in a desiccating environment. Therefore, while it is not at all impossible that animals moved from rivers on to land, it seems much more likely that they did so under conditions where some pre-adaptations to terrestrial life arose in water. As will be seen in succeeding sections, this is more likely in slow-flowing or stagnant water.

## 8.2.2 Lakes and swamps

Whereas rivers are characterized by continuous one-way flow, with the water having a short residence time at each place, the water in lakes has only a slow throughput, i.e. a long residence time. Lakes therefore show very different physical characteristics from rivers, because the deeper parts of them can become effectively isolated from the surface (e.g. Hutchinson, 1975). Deep currents may arise in lakes, but these are never so effective in mixing as the larger scale currents found at sea, which ensure oxygenation at almost all depths. In consequence, one of the major characteristics of lake water is a tendency towards de-oxygenation. In the temperate zone, a sharp

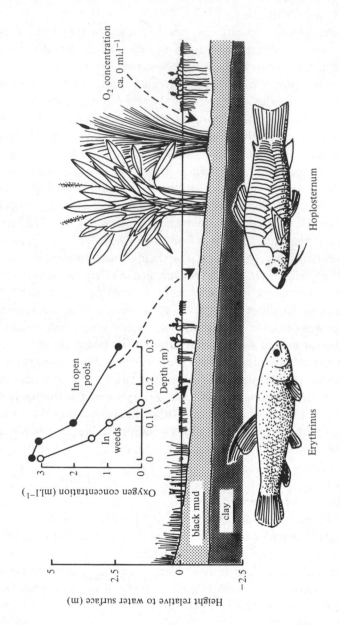

Fig. 8.1. Diagrammatic section through the edge of the Chaco in Paraguay. The floating plants are *Pistia, Azolla* and *Eichhornia*. Emergent vegetation at the edge consists of sedges, and the two taller plants in deeper water are *Thalia* (left) and *Typha* (right). Two of the commonest fishes in the swamp are *Erythrinus unitaeniatus* and *Hoplosternum litorale*. The graph shows that oxygen concentration is always low, even on the surface. After Carter & Beadle (1931a).

vertical gradient in temperature profile, or thermocline, is usually set up in summer, so that bottom waters may become de-oxygenated at that time, as organic matter, sedimenting into deeper water, utilizes the available oxygen. In the autumn, the equinoctial gales break up the thermocline and bottom waters become re-oxygenated. In tropical regions, however, with less seasonal change in heating, deeper lakes tend to de-oxygenation below the thermocline throughout the year. Shallow lakes, both tropical and temperate, may have no thermocline, because mixing induced by winds and by convection currents prevent its build-up. It might be thought that in this case de-oxygenation would not occur, but this is not true in the tropics: here, because of the high temperatures, the rate of decay of organic matter can be very high, and even areas of water less than 0.5 m deep can become entirely anoxic.

A classic study of such an environment was carried out by Carter & Beadle (1931a), who investigated the tropical swamps known as the Chaco, in Paraguay (Fig. 8.1). These swamps are dominated by the emergent macrophytes *Thalia* and *Typha*, both of which grow to a height of 5 m, and overshadow the water in many areas. Both towards the edge of the swamps, and in their centre, there are open spaces with floating vegetation such as *Pistia, Azolla* and *Eichhornia*. The water depth is seldom greater than 1 m because of the extreme flatness of the whole area, and in the summer, when evaporation is intense, the swamps may dry up completely on the surface. The swamp is underlain by an impervious clay which prevents water draining vertically downwards, and this clay never dries out totally.

The annual mean air temperature in the Chaco is 23.5 °C, with daily means varying from 18.5 in winter to 27.3 °C in summer. Maximum air temperatures may reach over 40 °C, and only occasionally drop below 10 °C. Rainfall does not show much seasonal variation: on average 800 mm fall in the summer six months, and 600 in the winter, but periods of rainfall are irregular, and there may occasionally be drought periods of as long as 6 months. Within the water of the swamp, Carter & Beadle showed that temperature extremes were not as great as in air, but surface maximum was still 40 °C, and bottom minimum as low as 15 °C. During summer, a distinct thermocline occurred, during the day, at a depth of about 10 cm. Although this disappeared at night, there was little mixing of surface and bottom water, presumably because of the thick layer of vegetation near the water surface, preventing access by wind. Consequently, although oxygen was sometimes found in the surface waters, the bottom waters were always anoxic. Even at the surface, oxygen levels seldom rose above 50% saturation. Only where surface weed was absent – allowing wind-driven mixing to occur – was oxygen found near the bottom of the swamp, and even then the water was only 20% saturated.

In these conditions, Carter & Beadle (1931a) concluded that the prime factor controlling animal life was the oxygen content of the water. Animals had the options of breathing dissolved oxygen in the very surface water layers, or of taking in air directly from above the surface. Since in many areas of water, there was no detectable dissolved oxygen in the surface water, the selection pressure for the evolution of air breathing must have been intense.

Thus it may be argued that, in fresh water, animal lines became best adapted for aerial breathing in swamp regions deprived of oxygen. This point can now be treated in some detail by reference to the evolution of air breathing in vertebrates.

### 8.3     The evolution of terrestrial vertebrates

Traditionally, terrestrial tetrapods have been envisaged as arising from freshwater ancestors during Upper Devonian times, and this viewpoint has been strengthened by the evidence discussed in Section 4.10. Recently, however, evidence has accumulated suggesting that the fish groups ancestral to tetrapods may have lived in shallow marine waters, or may have migrated back and forth between the sea and fresh waters, perhaps on a seasonal basis. Despite the fact that the fossil record for early vertebrates is probably more complete, and better studied, than that for any other animal group moving on to land, the discussion is still complicated by a lack of consensus about which fish groups are closest in relation to the early amphibians. While many authorities suggest a mono- or diphyletic origin of amphibians from the rhipidistian fishes of the Devonian, others have concluded that the lungfishes, and not the rhipidistians, may be the closest fish-like relatives of the tetrapods.

The various palaeontological arguments have been discussed by Little (1983), and these will not be repeated here, but a brief summary may help to visualize some of the possible relationships between early vertebrates. A classification of some of the vertebrates is given in Table 8.1 to provide an overall framework for the groups discussed. The two candidates for ancestry of the first amphibians are the rhipidistian fish, now all extinct, and the lungfish, still with living representatives. The majority view is that one group of rhipidistians, the Osteolepiformes, gave rise to the early amphibian labyrinthodonts (e.g. Jarvik, 1981; Panchen & Smithson, 1987; Carroll, 1988). Rosen *et al.* (1981), in contrast, have concluded that the lungfishes are the closest known relatives of the amphibians. Although the

Table 8.1 *A classification of some of the vertebrates mentioned in the text*

| | |
|---|---|
| Class Osteichthyes | |
| Subclass Acanthodii | |
| Subclass Actinopterygii | |
| Chondrostei | *Polypterus* |
| Holostei | *Amia* |
| Teleostei | *Clarias, Hoplosternum* |
| Subclass Crossopterygii | |
| Rhipidistians | Osteolepiformes, Porolepiformes |
| Coelacanthini | *Latimeria* |
| Subclass Dipnoi | *Neoceratodus, Lepidosiren* |
| Class Amphibia | |
| Subclass Labyrinthodontia | |
| Ichthyostegalia | *Ichthyostega* |
| Anthracosauria | *Seymouria* |
| Subclass Lepospondyli | |
| Microsauria | |
| Subclass Lissamphibia | |
| Urodela | salamanders, newts |
| Apoda | caecilians |
| Salientia | anurans – frogs and toads |
| Class Reptilia | |
| Subclass Anapsida | |
| Cotylosauria | captorhinomorphs |

labyrinthodonts are the earliest-known amphibians, these must have had precursors of which we have no knowledge. It now seems likely that it was these precursors, and not the labyrinthodonts themselves, that gave rise to the first reptiles, the captorhinomorphs. However, this overly simple summary hides the possibility of polyphyletic origins for several of the groups, and the whole situation can at present be said to be far from settled.

Rather than go into detail here, the controversy will be approached by considering the pressures that have led to air breathing and walking in the vertebrates, as opposed to the primitive characteristics of water-breathing and swimming. This will be done by looking in turn at the ecology of some present-day teleosts, then at the lung fish and finally at the fossil rhipidistians and the evidence from their evolutionary successors.

### 8.3.1 *Air-breathing teleosts and the swamp environment*

Although present-day teleosts are in no way ancestral to terrestrial vertebrates, the situations in which they have evolved similar attributes to those now found in semi-terrestrial and terrestrial vertebrates may provide circumstantial evidence for the selection pressures that brought about these attributes. In particular, this section will be concerned with the origins of air breathing, but some attention will be paid to the origins of walking.

The majority of fish, both in sea water and in fresh water, extract oxygen from the water by passing large volumes over the gill filaments. In the open sea this may be achieved mainly by the motion of the fish through the water, but benthic fish suck water into the branchial cavity and then expel it through the opercular opening. Fish that are exposed to the air occur primarily in two habitats: the marine littoral zone and inland freshwater swamps. The conditions of aerial exposure are quite different in these two regimes. The marine intertidal zone leaves animals emersed on a regular, usually semi-diurnal basis, governed by the tides: a fish left stranded at low tide will usually be covered 6 hours later. Inland freshwater swamps, on the other hand, have a regularity of water supply only on a seasonal basis: fish may be left out of water for months. On a more short-term basis, however, as described in Section 8.2.2, fish in these swamps may be covered by water but deprived of oxygen because of the high rate of organic decay.

If the distribution of specific adaptations to air breathing is summarized (Carter & Beadle, 1931b; Dehadrai & Tripathi, 1976), it is seen that air breathing in present-day fish is found mainly in tropical freshwater swamps and pools. Apart from a few isolated examples, only the gobioid fish such as *Periophthalmus* spp., discussed in Section 6.3.6, live on marine shores but show marked modifications for air breathing. Here we can now describe the life-styles of some of the freshwater fish, relating their respiratory mode to their environment.

One of the commonest fish found by Carter & Beadle (1931b) during their expedition to the Chaco swamps of Paraguay was *Hoplosternum litorale*. This catfish was found throughout the year in open water at the edge of the swamps, often in rapidly drying pools. It also travels from one pool to another overland, through the grass. It requires access to the water surface to survive, where it takes in air and swallows it. The air passes into a modified section of the intestine with very thin walls and a good blood supply, and oxygen is absorbed there. By use of this mechanism, it can live in water with very low oxygen content, a necessity as described in Section 8.2.2. Its breeding habits are also linked closely to the stressful character of the swamp environment. The fish begins to nest after rains fall in the early summer, by making a raft of plants at the water surface. The eggs are attached to the underside of this raft, and the whole mass is buoyed up by a secretion of foam. The foam probably provides an essential oxygen supply for the eggs, since the surrounding water quickly becomes completely anoxic.

In other continents, many freshwater teleosts show parallel physiological and behavioural adaptations to low oxygen level. The Indian species *Clarias batrachus*, for example (Dehadrai & Tripathi, 1976), lives in shallow water, and breathes air from the surface. Like *Hoplosternum, Clarias* emerges from the water at night, but in this case to feed on earthworms on the land. Its air-breathing adaptations are quite different from those of *Hoplosternum*: instead of taking air into the intestine, oxygen is absorbed through accessory respiratory organs in a specially developed cavity above the branchial cavity. Another Indian species, the climbing perch *Anabas testudineus*, also comes out on to land, and yet another, *Amphipnous cuchia*, spends the majority of its time out of water: the young even hatch in burrows above water level, and only enter the water after the gills have developed.

Several of these teleosts share the combination of ability to breathe air with the ability to move over the land surface out of water. This is not unique to freshwater teleosts, since the same combination is found in the marine *Periophthalmus*, but the ability to move using the fins for support seems to be more common in these freshwater fish than in marine teleosts in general. Some benthic marine forms like *Trigla*, the gurnards, use their pectoral fins for walking, and frog-fishes (Antennariidae) use the paired fins as limbs, but most of the marine littoral forms such as blennies and gobies use the tail and not the paired fins to hop from one pool to another (Marshall, 1965). In many cases, the fins of teleosts are used more as suckers to anchor the fish in position than as limbs, particularly in the intertidal zone where the action of waves tends to displace animals unable to attach to the substrate. Unfortunately, although a good deal is known about the respiratory adaptations of freshwater air-breathing fish, relatively little is known about their mechanisms of movement, or how this relates to their life-style and natural history. Perhaps with more information on this topic, it might be profitable to speculate further on the pressures that have brought about terrestrial walking in animals essentially adapted to fresh water.

In the cases of freshwater fish just discussed, there seems little doubt that the ability to breathe air has been developed because the oxygen levels in the water are unstable, and often low. No such examples of air breathing are found, for instance, in rapid-flowing streams where oxygen levels are constantly near saturation. Carter & Beadle (1931b) concluded that the breathing of air was not by itself related to any movement out of water, but was a necessity brought about by lowered oxygen levels, particularly in shallow, slow-flowing water in the tropics. While such a generalization may not necessarily have been true of processes operating in the Devonian, it is at least interesting that air breathing is now much more common in tropical fresh water than in the marine intertidal zone. This is a point to return to in discussion of air breathing in rhipidistians (8.3.3).

### 8.3.2    *Lungfishes and the swamp environment*

Three genera of lungfish are extant today: *Neoceratodus* in Australian rivers, *Lepidosiren* in swamps of South America, and *Protopterus* in African swamps. Of these, *Neoceratodus* is thought to be the most primitive. It has fully functional gills, lives in fully oxygenated waters, and can survive only poorly out of water. Its fins have an axial skeleton like rhipidistians, and have not been reduced to thin filaments as in other genera. Although they allow the fish to walk under water, they cannot support the body on land. Nevertheless, *Neoceratodus* has a functional lung, connected to the pharynx, and, when it needs more oxygen than the gills can supply, it breathes air from the water surface. It has been postulated that, with this lifestyle, there is sufficient selection pressure to *maintain* the lung, but not enough to explain its origin.

For a situation in which a lung might have arisen, we must turn to *Lepidosiren*. This fish lives in the same swamps of the Chaco harbouring the teleost *Hoplosternum* (8.3.1). *Lepidosiren paradoxa* has paired lungs, and unlike *Neoceratodus* is an obligate air breather, coming to the surface as soon as it hatches from the egg. This ability is essential, since, unlike the situation for *Neoceratodus*, oxygen levels in the water are often very low, as discussed above. Aerial respiration is also essential during the dry season, when *Lepidosiren* burrows down into the mud and aestivates at the bottom of the burrow. Activity is almost negligible during this period, but the fish become active again after the first rains.

A similar pattern is seen in the African species *Protopterus annectens*, but here the swamps dry up more completely than those in South America. *Protopterus* burrows into the mud and forms a mucous cocoon around itself at the onset of the dry season (Johnels & Svensson, 1954). It aestivates for periods of up to 9 months, and during this time it breathes air, using its double lungs. As soon as water returns to the area, the lungfish emerge, start feeding, and prepare for breeding. Even then, adults obtain most of their oxygen from the air, not from the water, although the pattern is different in juveniles (Fig. 8.2). They can swim in open water, but also walk on the bottom, using their fins as supports and as sensory probes for benthic mollusc and crustacean food. Breeding nests consist of horizontal burrows into the banks in shallow water. After the female has laid her eggs,

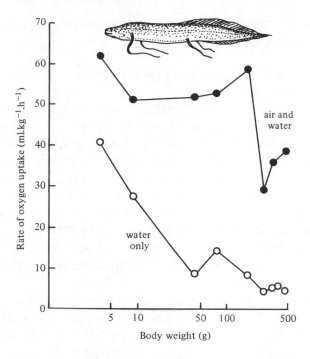

Fig. 8.2. Oxygen uptake in the African lungfish *Protopterus amphibius*. When allowed access to air, uptake rate was greater, especially in larger fish. In adults, almost all uptake came from air, and the gills were responsible for less than 20% of the total uptake. After Johansen, Lomholt & Maloiy (1976).

it is the male that guards the nest and keeps a flow of water through the tunnel. The young remain in the nest until they are 30–35 mm long, and then leave to begin life on their own.

In summary, *Lepidosiren* and *Protopterus* appear to be excellently adapted for life in tropical swamps where dissolved oxygen is scarce and water supply is seasonal. *Neoceratodus*, on the other hand, is found in conditions of much greater constancy. It is believed, though by no means proved, that lungs originated in the ancestors of lungfish under the pressure for air breathing in tropical swamps, and that from these same ancestors evolved the lungs of the tetrapods. The ability of lungfish to walk, albeit only under water, may well have been a response to conditions where water supply was uncertain, and where movement in shallow water and water films was essential as a preliminary to aestivation. Alternatively, limbs could have been derived in shallow water with heavy weed growth, where the ability to push through and round aquatic algae would have been considerably enhanced by the use of flippers.

### 8.3.3     Palaeohistory: the ecology of rhipidistians and coelacanths and the origins of the tetrapods

Despite the circumstantial evidence just considered, it must be made quite clear that views about the environment in which the early tetrapods actually evolved are diverse. This uncertainty applies also to the environment of the early aquatic rhipidistians – the lobe-finned fishes of the Devonian. Initially, the majority view placed these in freshwater regions, but further consideration has suggested that many of these fossil forms may have been marine, or may have moved seasonally from marine to freshwater systems (Thomson, 1980; Bray, 1985). Thomson (1980), particularly, has made a good case for the fact that air breathing in rhipidistians *could* have evolved in salt water, although the view taken here favours a freshwater origin.

The two major groups of rhipidistian fish were the Porolepiformes and the Osteolepiformes. Both had paired fins with axial skeletons, and are known as the lobe-finned fish. They also had a braincase divided by an intracranial hinge, allowing considerable movement between each part. They were predators, and the general feeling is that they were able to breathe air, through the internal nostril or choana. The fins would never have allowed the fish to walk on land, but aided them in 'walking' under water, in a fashion similar to the present-day lungfish *Neoceratodus*.

The Porolepiformes come from both marine and freshwater deposits of the Devonian. As Thomson (1980) and Bray (1985) have pointed out, it is not at all easy to decide upon the living conditions of such fossil forms, because of the complex geography of Devonian fossiliferous areas (see Section 3.1.3). While some freshwater areas were without direct contact with the sea, others merged with the marine zone through deltaic conditions, and the marine and freshwater influences there probably fluctuated seasonally. The possible movement of the skeletons of freshwater fish to the marine littoral zone under these circumstances also complicates the issue. It is possible that the early Porolepiformes gave rise to specific groups of modern amphibians. Jarvik (1980) has suggested that they were ancestral to present-day urodeles, although other authors (e.g. Jurgens, 1971) have refuted this idea.

Whether the Porolepiformes evolved in the sea or in fresh water, they are probably closely related to another crossopterygian group, the coelacanths (Andrews, 1973). This group, represented by the extant *Latimeria chalumnae*, first appeared in the Devonian fossil record in marine sediments (Thomson, 1980), but had lobe fins and an air bladder. Whether they or their ancestors with these features evolved in the sea or fresh water is unknown, although as argued above, the selection pressure for both accessory air breathing and for the use of the fins in walking is easier to see in fresh water.

The Osteolepiformes first appeared in deposits laid down in mixed fresh and sea water, and again it is difficult to be certain of the conditions in which they evolved. Bray (1985) suggested that in Devonian times the so-called 'fresh' waters would have had a higher ion content than now, and that this would have facilitated the movement of fish between environments in the different seasons: marine fish, particularly, would have been able to move into fresh water more easily. Thomson's (1980) view that many rhipidistians might have been anadromous – living in the sea but entering fresh waters to breed – may tie in here. On this basis, however, it is still hard to see what the selection pressures for air breathing would have been. Bray has suggested that marine fish invading continental lakes and rivers in the dry season would have been driven on to land by the dilution that followed seasonal rains; but such a suggestion is far removed from any response predictable from our knowledge of the physiology and ecology of present-day fishes, and needs supporting evidence before being considered further.

Osteolepiformes are widely regarded as the most likely ancestors of the amphibians. Jarvik (1980) has suggested that, of the modern amphibians, they may have given rise only to the anurans, but this is a minority view. Possible interrelationships of the tetrapods and various fish groups are indicated in Fig. 8.3. The most unsatisfactory gap here is the relation between Osteolepiformes and the first Devonian amphibian. The earliest of these that have been discovered were the labyrinthodonts known as the Ichthyostegalia, but these had advanced far beyond rhipidistian form. They had already lost the rhipidistian hinge in the brain case. The limbs terminated in individual digits instead of being flippers. The pectoral girdle had lost its connection with the skull and was strongly connected to the ribs and vertebral column. The pelvic girdle was larger, and connected also to the vertebrae. Genera such as *Ichthyostega* were probably amphibious, able to swim or crawl in shallow water, but also able to crawl over the land surface. *Ichthyostega* almost certainly breathed air through a lung or lungs, but could also have had gills, as do the lungfish. A theory put forward by Romer (1958) suggested that freshwater pools of the Devonian would have been temporary, and as they dried up there would have been a selective

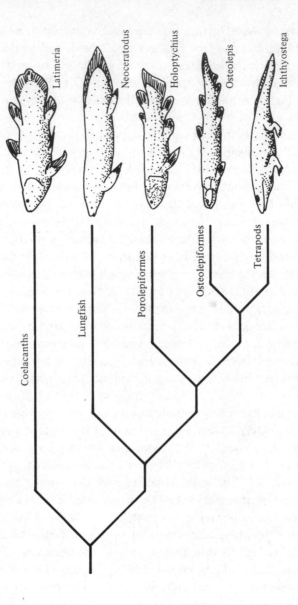

Fig. 8.3. Possible interrelationships of tetrapods and fishes, based on Panchen & Smithson (1987). In this view the Osteolepiformes are most closely related to the tetrapods, but other schemes such as those of Gardiner (1980) and Rosen *et al.* (1981) put the Dipnoi nearest to the tetrapods.

advantage in being able to crawl to the next pool. However, the idea of an environment in which the pools were surrounded by relatively arid areas does not fit with our conceptions of the early amphibians as animals that easily lost water (Inger, 1957; Thomson, 1980), and, although the swamps may have resembled present-day swamps in the tropics, seasonal drought with its concomitant selection for aestivation is not likely to have been one of their properties. Current attitudes seem to be more that the early amphibians lived in environments of relatively high and constant humidity: swamps that provided food and water all the year round rather than the seasonal environments of the present-day South American and African lungfish. In other words, the ability to aestivate, while it has been evident in lungfish since Devonian times, has not itself been a precursor of terrestrial life in the tetrapods, but a specialist adaptation of particular groups.

Although there was at one time a view that the labryinthodont amphibians, particularly the genus *Seymouria*, gave rise in turn to the early reptiles (Watson, 1951), this now seems unlikely. The earliest definite reptiles were the captorhinomorphs from the mid Carboniferous. It has been strongly argued (Panchen, 1980) that these are unlikely to have been derived from the groups containing *Ichthyostega* or *Seymouria*, because, while the early reptiles had no indentation at the back of the skull for the middle ear (an 'otic notch'), all the labyrinthodonts possessed such a notch. Since later reptiles also have a notch, the evolutionary history of the reptiles would therefore have involved first the development of the notch, then its regression, and finally its re-development. There is no evidence for this complex series, and further studies by Lombard & Bolt (1979) and Carroll (1980) have pointed to a divergence of evolutionary lines before the origin of the captorhinomorphs. In other words, the early reptiles may have been derived not from labyrinthodonts, but from earlier 'proto'-tetrapods. This may mean that reptiles as well as recent amphibians evolved directly from rhipidistian (or possibly dipnoan) ancestors.

Whatever their origin, captorhinomorphs were the first reptiles to survive as fossils. Early forms were small, and it has been argued by Carroll (1970) that they must have evolved from small amphibian ancestors, maybe no bigger than 10 cm from snout to vent. Carroll's argument was based on the knowledge that amphibian eggs require direct diffusion to supply them with oxygen, and at present this restricts the largest amphibian egg to about 9 mm in diameter. Since there is also a relationship between egg size and adult size in amphibians, it is likely that early terrestrial amphibians were also small. The idea of the small size of early reptiles also fits well with our conception of the kind of habitat in which captorhinomorphs originated:

with a large surface area to volume ratio even in adults, early captor-hinomorphs must have lived in humid micro-habitats. Only as the eggs began to develop accessory breathing organs, and could therefore become larger, could the adults increase in size. This later type of egg, the amniote egg, was the major advance that allowed the vertebrate classes to expand into the great diversity of terrestrial niches. Whether or not such eggs evolved more than once is still disputed: several independent origins could mean that mammals and reptiles evolved entirely separately, or even that the reptiles are polyphyletic (Stahl, 1974; Kemp, 1980). However that may be, the protection of the embryo that characterizes reptiles, birds and mammals, together with the development of waterproofing in the adult, has provided a very successful basic design from which these groups have subsequently radiated, to inhabit even the driest environments on land.

### 8.4    Prosobranch gastropod molluscs: Viviparidae, Ampullariidae and Cyclophoridae

If adaptation to the conditions in freshwater swamps has allowed the invasion of land by vertebrates, it would be surprising if other swamp-living groups had not also to some extent become pre-adapted for land life. Of the molluscs, a prime example is provided by the prosobranch family Ampullariidae. Some genera in this family, for instance *Turbinicola*, are found in fast-flowing streams, but the majority, including *Pomacea* and *Pila*, are restricted to freshwater swamps.

*Pomacea depressa*, like most other ampullariids, has a globose shell, with a tightly fitting operculum. It lives in the Everglades swamps in Florida, where temperatures are high, evaporation rate is rapid and the supply of organic matter from the growth of emergent vegetation is abundant. *P. depressa* possesses both a gill and a lung: the mantle cavity is divided in two, and the left side forms a well-vascularized lung, with the anterior edge formed into a tubular siphon that can be extended to reach the water surface. Particularly in de-oxygenated water, the snails come to the surface and draw air into the lung by contractions of the body muscles (McClary, 1964). Although the structures involved are quite dissimilar from those used by the amphibious vertebrates, the functional parallel with lungfish and air-breathing teleosts is very close.

*Pomacea* can also move some distance out of water. It does not do this to feed, however, but to lay its eggs. It climbs up the stems of reeds or grasses, and lays the eggs in batches. The mucous coats around the eggs harden as they dry, forming a tough impermeable barrier. When the young hatch, they penetrate this barrier and drop back into the water. The functional significance of laying eggs out of water is not clear, but may well represent a method of keeping the eggs away from anoxic conditions, and also out of the range of aquatic predators. It must be said also, however, that the movement of the snails even to the water surface is hazardous for them, as they are preyed on by a variety of birds. In Central and South America, species of *Pomacea* form a large proportion of the diet of one predator, the snail kite *Rostrhamus sociabilis*. This bird captures one snail at a time and retreats to a perch to extract the living soft parts from the shell (Bourne, 1985).

A further parallel to the swamp-living fish is provided by the ability of *Pomacea* to aestivate in the dry season (Little, 1968). After burrowing in the mud, it closes the aperture of the shell with its horny operculum, and also reduces its rate of metabolism. Temperature relationships of *P. urceus* during aestivation are shown in Fig. 8.4. Aestivation can last for months, but the snails rapidly become active again when covered by water. A similar ability is found in African genera such as *Pila*. These 'apple snails' have a similar life-style to *Pomacea*, being active in the wet season and aestivating when the swamps dry up. They emerge from aestivation partly from the stimulus of water cover, and partly because of the mechanical disturbance.

Ampullariids have two sets of relatives, and with them comprise the superfamily Architaenioglossa. The abilities of this group in terms of osmoregulation and water balance have been discussed in Section 4.4.2. In fresh water, ampullariids are related to the family Viviparidae – snails characteristic of slow-moving water in muddy rivers and lakes, but without the ability to crawl out of water or to aestivate. Viviparids brood their young in a uterus, and the young then leave the parents as miniature adults. There are no close marine relatives of viviparids, but they are not highly differentiated from marine littoral mesogastropods, except that they have ion-regulatory mechanisms typical of freshwater invertebrates. It can be surmised that snails not unlike present-day littorinids moved through estuaries to fresh water, and from these evolved viviparids in rivers and lakes, and ampullariids in swamps.

The second relatives of ampullariids are the Cyclophoridae – a family, or more probably a series of families – found on land. Cyclophorids have a variety of shell forms, and are found all round the world in tropical and subtropical areas. Most are inhabitants of rainforest leaf litter, but some are arboreal, and others occur in rocky limestone districts. The mantle cavity of cyclophorids consists of a lung, and the snails breathe air like the more common pulmonate land snails. Aestivation is probably common in the dry season outside the forests, and all cyclophorids, including the forest

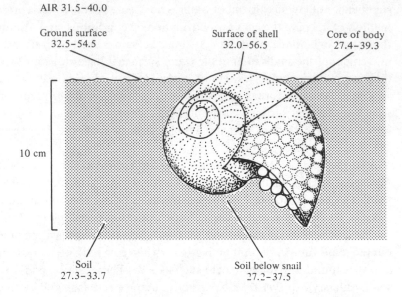

AIR 31.5–40.0

Ground surface
32.5–54.5

Surface of shell
32.0–56.5

Core of body
27.4–39.3

10 cm

Soil
27.3–33.7

Soil below snail
27.2–37.5

Fig. 8.4. The temperature relationships of an aestivating prosobranch snail from freshwater swamps, *Pomacea urceus*. Although the shell surface becomes as hot as the ground surface, the body of the snail and the eggs, buried in the ground, remain below 40°C. Compare this with the situation for a desert pulmonate on the surface of the ground (Fig. 2.7). After Burky, Pacheco & Pereyra (1972).

species, are dependent upon humid conditions for activity. They are not conspicuous snails, since most of them eat dead and decaying leaves and not living vegetation, and tend to remain partly camouflaged on the forest floor. They are in general adapted to conditions of constant high humidity, and apart from the adoption of air breathing they have few features that differentiate them from aquatic snails.

It is unlikely that cyclophorids have evolved directly from snails like the present-day ampullariids, which are specialized for life in modern swamps. More acceptable is the idea that they evolved from freshwater ancestors with some of the characteristics of viviparids and a tendency towards air breathing. The selection pressure for this air breathing may well have been a lack of dissolved oxygen, as suggested for the rhipidistians and lungfish, and exemplified by the ampullariids. As with the early tetrapods, life on land must have originated when terrestrial micro-climates with constant humidity were available in immediate proximity to freshwater swamps. The difference from the vertebrate situation is that cyclophorids have never developed either a protected egg or an epidermis impermeable to water. Unlike the vertebrates, therefore, cyclophorid snails have remained restricted to tropical rain forests, which provide them with one of the most constant terrestrial environments available.

## 8.5    Prosobranch gastropods: Neritidae and Helicinidae

Despite the fact that there are approximately 100 marine families of prosobranch gastropods at the present, there are only about 10 freshwater families (Little, 1985). Apart from the Viviparidae and Ampullariidae discussed in Section 8.4, only the Neritidae are related to any considerable group of terrestrial snails, and the other families will not be considered here.

Freshwater neritids are undoubtedly derived from marine forms. Indeed, the family is characteristic of tropical and subtropical shores worldwide, from rocky substrates to mangrove swamps (see 6.3.2). In fresh water, neritids are usually found on rocks, where current flow is fast and there is little likelihood of low oxygen concentrations. *Theodoxus fluviatilis*, for instance, creeps over stones, wood and aquatic plants in European streams, browsing on detritus and epilithic algae. It is usually restricted to rivers, lakes and canals with a high calcium concentration, but in some areas of the Baltic it occurs in brackish water (Fretter & Graham, 1962).

Within the family Neritidae, one example appears to be completely terrestrial: *Neritodryas*, found in the Indo-Pacific, lives on bushes and trees, often far from water (Cooke, 1895). Unfortunately, there have been no recent studies on this fascinating group. It seems unlikely, however, that the evolution of such genera would have taken place in swamps, since neritids in general show little modification for life in de-oxygenated waters. In Section 6.3.2 it was suggested that some neritids might have followed the mangrove habitat into fresh water, and as an extension of this hypothesis, it may be added that freshwater neritids could well have invaded the land by retaining an arboreal habit, and moving from trees living in fresh water to inland forests.

This idea may be supported by consideration of a family of terrestrial snails, the Helicinidae. This family may be closely related to the Neritidae, but some authors such as Thompson (1980) have cautioned that helicinids should constitute a separate superfamily. Whether or not they are closely related to neritids, it has been proposed that helicinids may also have a freshwater ancestry (Little, 1972). Solem & Yochelson (1979), however, have shown that, as yet, the earliest terrestrial helicinid fossils come from the Upper Carboniferous, pre-dating the earliest freshwater neritids (found

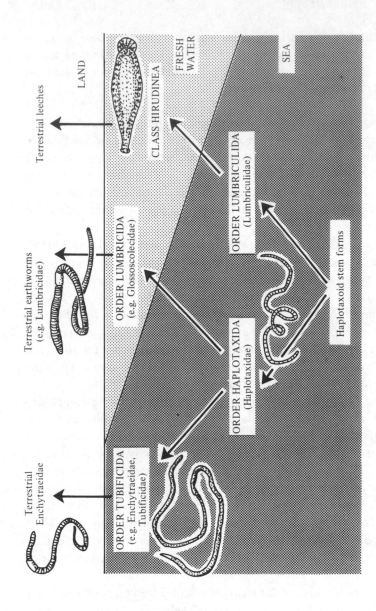

Fig. 8.5. Possible evolutionary routes of annelids on to land. Some groups, including the Enchytraeidae, may have moved directly to land from the sea. Others, such as those giving rise to present-day earthworms and leeches, invaded fresh waters and only then moved on to land. Taxonomic relationships after Brinkhurst (1982), but environmental routes original.

in the Jurassic) by a considerable period. If these records genuinely reflect the times at which the groups appeared in these habitats, ancestry of helicinids from freshwater neritids would obviously have been impossible. Three points may be made here. First, the helicinid described by Solem & Yochelson may in fact not belong to this family. Secondly, the record for freshwater neritids may well be incomplete. Thirdly, if Thompson (1980) is correct, helicinids evolved not from neritids but from some other, unknown, ancestors. Although the observations of Solem & Yochelson are very valuable, they do not therefore preclude the possibility of a freshwater origin for helicinids.

Helicinids are common on islands and on coastal fringes of larger landmasses in the tropics, particularly in the Caribbean and the Indo-Pacific. Most helicinids are tree-dwellers, and are active only in humid or wet conditions. Some are restricted to leaves, but many are found on the trunks of trees, grazing on algal and detrital films. Occasionally they are also found on limestone rocks, and many of the species are restricted to areas where there is a good calcium supply. Helicinids have a vascularized mantle cavity acting as a lung, forming a striking parallel with the cyclophorid land snails discussed in Section 8.4. Most have an operculum with which the shell aperture can be closed, but Thompson (1980) regarded this as a secondary development.

The habits of helicinids have been very little studied, but their distribution on trees and rocks probably reflects their ancestry on hard substrates in fresh water. It is to be hoped that further study of their ecology and physiology will throw more light on the ways in which terrestrial gastropods have utilized the freshwater environment as a route on to land.

## 8.6 The cryptozoic habitats: oligochaetes and leeches

Up to this point we have considered in this chapter only the animals living outside the soil matrix. As in marine systems, however, many groups live within the substrate, either in burrows or within the interstitial spaces, and these include the annelids. The alternative hypotheses of a direct (marine) origin and an indirect (freshwater) origin for terrestrial oligochaetes have been discussed in Sections 4.9.2 and 5.2.6. In Section 5.2.6 it was suggested that, while terrestrial enchytraeids probably evolved directly from marine ancestors, earthworms are more likely to have originated in fresh water. A possible evolutionary sequence showing these two lines is presented in Fig. 8.5. The following section is devoted to the earthworms and their origin, and to the terrestrial leeches which have not so far been discussed.

### 8.6.1     The earthworms

There are three major families of earthworms at the present day. The largest is the Megascolecidae, containing the genera *Pheretima* in India and *Megascolecides* in Australia. The megascolecids are not found in northern Europe or North America, and because of this distribution have not been intensively studied. The Glossoscolecidae are found mostly in the tropics and subtropics, particularly in South America, while the genus *Alma* is found in the swamps of east Africa. The family that has elicited most study is the Lumbricidae, typical of temperate regions, and containing such widely known species as *Lumbricus terrestris, Allolobophora caliginosa* and *Eisenia foetida*. It is the study of these species that has emphasized the ecological importance of earthworms in digesting dead angiosperm material and mixing it with soil particles. Because earthworms live in burrows, drag food material into them, and then defaecate at the soil surface, they are responsible, as shown by Darwin (1881), for a great deal of soil mixing and aeration. In spite of their cryptozoic habit, they are immensely important in terrestrial ecosystems, as will be emphasized in Chapter 10.

Lumbricids are active throughout the year in soils with a wide range of water contents. During the summer dehydration period, many species tend to burrow deeply, but others can remain active near the surface (Edwards & Lofty, 1972). Glossoscolecids, in contrast, require high rainfall, and this probably restricts them to specific areas of the tropics. Megascolecids show a variety of responses to moisture. Genera such as *Pheretima* can live in dry habitats and can remain active even in the dry season. Partly, at least, this is because their nephridia provide a good water-conserving system. *Eutyphoeus*, which has no specialized system for water retention, lives only in wet soil and burrows deeply in the dry season (Bahl, 1947).

The species that are found in fresh water and in swamps are thought to have re-invaded these environments from terrestrial soils (Stephenson, 1930). For example, the glossoscolecid *Alma emini* lives in the papyrus swamps of east Africa (Beadle, 1957). Like the swamps of South America, described in Section 8.2.2, the waters in these papyrus swamps are often anoxic. Conditions of stagnation in the water trigger the formation by *A. emini* of a temporary 'lung' at the hind end of the body. This lung area is formed by the sides of the body contracting to form a semi-enclosed area around a well-vascularized surface. When it is protruded into the air, this surface can take up 50% of the oxygen normally taken up through the whole body surface under water (Mangum, Lykkeboe & Johansen, 1975).

The original habitats of the freshwater ancestors of the earthworms – the 'proto-lumbricids' – are still obscure. The ability of modern earthworms to osmoregulate in dilute solutions, and their re-invasion of fresh waters on many occasions, make a freshwater origin most likely. The whole oligochaete organization is adapted for burrowing, and it is probable that this organization has remained unchanged while oligochaete ancestors moved from the sea to estuaries to fresh water. The type of freshwater habitat from which burrowing earthworms emerged into cryptozoic terrestrial habitats remains uncertain.

### 8.6.2  The leeches

Terrestrial leeches are much more restricted in their distribution than the earthworms. The major family, Haemadipsidae, is characteristic of the tropics, and haemadipsids are common in India, Indonesia, Australia, Madagascar, South America and some Pacific islands. Other families are also found in north-east Europe and in South America. Unlike earthworms, terrestrial leeches do not live in the soil, but are found in trees and shrubs and in leaf litter below them. They feed by sucking blood from rainforest mammals. Most respond to shadows and to warmth, and using these cues as well as chemical senses, they attach to passing animals. Relatively little research has been done on terrestrial leeches, and they offer a fruitful field for ecological and physiological study.

In temperate zones, other leech families are amphibious, and these may give some guide as to how the truly terrestrial families evolved. The family Erpobdellidae, for example, is predominantly found in the benthos of lakes and rivers, often where the bottom is muddy and with a high organic content. The genus *Trocheta* may also be found in terrestrial soils, burrowing and feeding on earthworms and slugs. A parallel is provided by the family Hirudidae. Most of the members of this family are found in fresh water, including the medicinal leech, *Hirudo medicinalis*. *Haemopis*, however, can also be found in soils. This amphibious habit of some of the primarily freshwater genera strongly suggests that the terrestrial groups could have originated from freshwater ancestors. The suggestion is supported by the point that there are relatively few marine leeches: about 75% of present day genera are found on land or in fresh water (Mann, 1962). The burrowing habit is widespread in aquatic leeches, and their movement from aquatic to terrestrial soils seems most likely. The subsequent abandonment of soils for leaf litter and arboreal habitats may be linked with the change from a diet of invertebrate prey to the habit of sucking blood from mammals; but it must also be pointed out that many aquatic leeches are bloodsuckers, and it is possible that this is a primitive lifestyle for terrestrial leeches. More research on the Haemadipsidae might well provide insights into their evolutionary history.

## 8.7    The cryptozoic habitats: flatworms (Turbellaria)

In temperate zones, terrestrial flatworms are relatively rare. In tropical rain forest, however, triclads can usually be found in leaf litter or under logs on the forest floor. Some of these triclads are small enough to be confused with terrestrial nemertines, but others have striking colouring, and may reach lengths of 60 cm. They are predators, eating slow-moving invertebrates such as earthworms, nematodes and snails. Activity is restricted to periods of high humidity, mostly at night, and during the day they retire to dark, humid refuges.

There are three major families of terrestrial triclads, grouped in the suborder Terricola. The most familiar in temperate zones is the family Rhynchodemidae. *Microplana terrestris* and *Rhynchodemus bilineatus*, for example, are found in humid situations in Europe. They are more cylindrical in cross-section than most flatworms, and have a specialized creeping sole on which they move. In south-west England they share retreats under rocks and logs with the carnivorous slug *Testacella*, and may have a similar lifestyle. Members of the family Geoplanidae are mostly South American, but some are now found in California. They were probably introduced with plants, and have become established as part of the cryptozoic fauna of gardens. The Bipaliidae contains the most distinctive species. These have a semi-circular shaped head bordered with numerous eyes, and supported on a narrower neck. *Bipalium kewense*, now common as an introduced species in southern parts of North America and in Europe, is an earthworm predator. Its natural habitat is the rainforest of Indo-China, but it is now sufficiently numerous in America to be a pest in earthworm farms.

Although the central habitat of Terricola appears to lie in tropical rain forest, some species are found in much more rigorous environments. These are normally refuges such as the undersides of stones or logs in grassland, but some species are also found in soil. When the surface layers dry up, these species migrate downwards, just as do several species of earthworms. All the species outside forests probably utilize isolated refuges during the day, and emerge to feed at night when conditions are humid and cool. Like all cryptozoic species, behavioural mechanisms rather than physiological ones ensure their success on land.

The relative success of terrestrial flatworms can be gauged by the fact that there are estimated to be 500 species at the present day – compared with about 20 species of terrestrial nemertine. The origins of the Terricola can be considered by comparing them with freshwater triclads (Paludicola) and marine triclads (Maricola). A few species of Maricola live in the rocky intertidal zone in crevices, and some specialized forms such as *Procerodes littoralis* inhabit shingle substrates where freshwater streams run into the sea – the type of environment described in Section 7.3 (see Fig. 7.2). *P. littoralis* has to withstand rapid changes from fresh to sea water. Using specialized cells in the gut it is able to excrete large quantities of water entering osmotically. This efficient system does not, however, appear to be ancestral to any regulatory system in the Terricola, all of which use a well-developed network of protonephridia for water balance. The freshwater Paludicola, in contrast, may well provide an organization ancestral to the Terricola, since they also use an extensive protonephridial system. They are abundant in fresh waters all over the world, where they form an integral part of many ecosystems as predators of oligochaetes, molluscs and crustaceans. In some ways they provide an ecological parallel with freshwater leeches, and there may be substantial competition between leeches and flatworms in some environments (Young, 1981). As shown in Fig. 8.6, lake triclads are normally restricted to waters with a high calcium content, but this is an indirect relationship governed by rates of production and hence availability of prey (Reynoldson, 1966). Competition for food is one of the major influences on the distribution of lake species (Reynoldson, 1983). Physical and chemical factors may be more directly relevant in controlling the distribution of stream species (Reynoldson, 1981; 1983).

The idea of a freshwater ancestry for Terricola is not unchallenged. In the past, they have been considered closer in organization to the Maricola than to the Paludicola, and Ball (1981) has pointed out that the ecological criteria separating Maricola from Paludicola are not entirely clear-cut, since some species of Maricola have now been found in fresh water. There is also a possibility that the Paludicola are not a monophyletic group, and that two groups of freshwater triclads and the Maricola may all have had independent origins from marine ancestors. Until further work has established relationships within the triclads, the origins of the Terricola must remain under debate.

## 8.8    Burrowers: the crayfishes

In the marine intertidal zone, many groups of animals depend upon their burrows for protection against predators and for provision of an appropriate micro-climate: in mudflats, several families of crabs (5.1.4) and

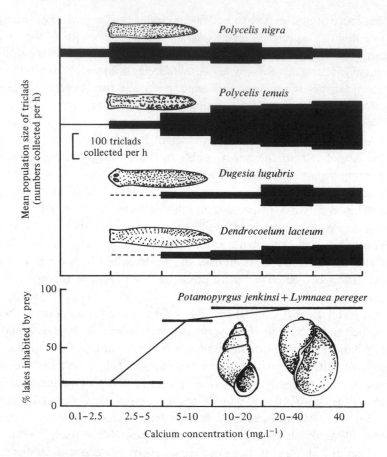

Fig. 8.6. Abundance of triclad flatworms and their prey in lakes of differing
calcium concentration. The top diagrams show the abundance of four species
of triclad as the average of the number collected in one hour. The lower
diagram shows percentage of lakes inhabited by molluscan prey (*Potamopyrgus
jenkinsi* and *Lymnaea pereger*). Both triclads and molluscs are commoner when
calcium levels are high. Invasion of land from fresh waters may have been
more prevalent where calcium was high and animal diversity was large. After
Reynoldson (1966).

many polychaetes burrow, and in sands there are several burrowing
isopods, amphipods and crabs (5.2.8 and 5.2.10). Apart from the
polychaetes, not considered in this book, the predominant among these are
crustaceans. The same dominance of fossorial habits by crustaceans applies
to the situation in fresh water. Crayfishes and crabs will be considered in
this and the following section.

Burrowing crayfish are widely distributed in North America and

Australia. The American species belong to the family Cambaridae, and *Procambarus gracilis* is an example of a species that burrows in ditches, temporary pools and wet meadows (Hobbs & Rewolinski, 1985). Its burrows are as much as 2 m deep, extending well below the water table which is only about 0.5 m below the ground surface. Each burrow is therefore water-filled at the bottom. Individual burrows usually contain only one crayfish, so that this species presumably lives a solitary life except at the breeding time.

Australian crayfish constitute the family Parastacidae. They have similar habits to the American species, but show a wider ecological range (Horwitz & Richardson, 1986). Some species burrow in or near permanent water bodies such as lakes and rivers. These are essentially fully aquatic animals, and probably represent a primitive stock from which more terrestrial species have evolved. A second group of crayfishes forms burrows in terrestrial soil that reach down to the water table. These are equivalent in situation to the American *P. gracilis*, but the burrows may contain both a male and a female, as well as occasional juveniles. Dispersal occurs when the level of groundwater rises and floods the burrows. The third group forms burrows that are independent of the water table. These crayfish, of which *Engaeus tuberculatus* is a good example, live on hill slopes in areas of high rainfall, and water is collected in the burrows from surface run-off (Fig. 8.7). Like the other species, therefore, there is a water-filled chamber at the bottom of the burrow, but the source of this water is quite different. Each burrow is more complex than the simple vertical burrows of the more aquatic forms, and typically has a central chamber with several tunnels connecting this to the surface. Within the burrow system are usually found members of several generations, forming a large family group (Horwitz, Richardson & Boulton, 1985). The occurrence of many individuals in one burrow is otherwise rare in crayfish, because most are aggressive and strongly territorial. Presumably the aggressive behaviour has been reduced in the terrestrial forms because they often live far from water, and the opportunities for dispersal are rare: family life is a necessity.

There are few observations on movement of crayfish overland. Dispersal of the more terrestrial species must occur in this way, possibly at night. Certainly, feeding of some species takes place on land, again probably at night. Crayfish eat a mixture of plant materials and occasional invertebrates, but plants form the major part of the diet. In some areas, the more terrestrial species can cause agricultural problems: the large burrows with central chambers tend to collapse, forming a hazard for cattle and horses.

There seems little doubt about the origins of the terrestrial burrowing

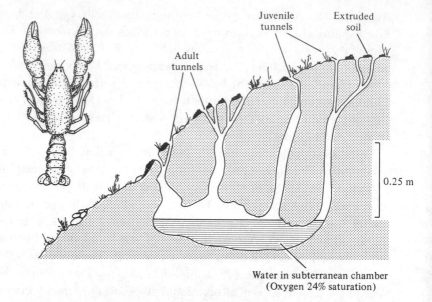

Fig. 8.7. The burrow system of a terrestrial Australian crayfish, *Engaeus tuberculatus*. The water in the burrow is independent of the water table, and is probably supplied from surface run-off. It is not well oxygenated, presumably because it is stagnant apart from any activity of the crayfish. After Horwitz, Richardson & Boulton (1985).

crayfish: they have been derived from a freshwater stock, and have been surprisingly successful on land. Since they can not themselves withstand a great deal of water loss, this success must be ascribed primarily to their use of burrows.

## 8.9    Burrowers: the freshwater crabs

Surprisingly little is known of the ecology and natural history of the crabs found in and near fresh water. There are maybe 11 families of crabs typical of fresh water, as well as some genera within the family Grapsidae that have moved into rivers and even on to the land. Two species have already been discussed in Section 4.7.3 in terms of their osmoregul-ation: *Potamon niloticus* from the Middle East, and *Holthuisana transversa*, from Australia; but other species are common in South America, India, Indonesia and the Far East. Many of these are restricted to an almost aquatic life in rivers and streams, but even some of the grapsids, such as *Eriocheir sinensis*, the Chinese mitten crab, can move overland when migrating up rivers. The burrowing activities of *E. sinensis* are currently a source of anxiety to river authorities, because this species is now common in Europe, and the burrows can undermine river banks. In tropical Asia, similar problems are caused by species of *Paratelphusa*, which burrow into the 'bunds' separating paddy fields.

Burrows are used as refuges, especially during the day, and the habits of *Paratelphusa* can be used as an example. Some species of *Paratelphusa* are active as predators during the day, but in general the crabs emerge only at night in wet conditions. The burrows reach down to the water table, and the crabs use the water in the burrow to keep their pleopods moist. Reproduction occurs within the confines of the burrow, and the female emerges only when the young have hatched and are attached to her pleopods. These young are then released into fresh water, where they lead an entirely aquatic life before returning to the land as adults.

A second example is provided by *Holthuisana transversa*, an inhabitant of north-eastern Australia (MacMillen & Greenaway, 1978; Greenaway & MacMillen, 1978). This crab differs from all other known terrestrial crabs in that it lives in areas classified as desert. In these regions, rainfall may be less than 200 mm/year. *H. transversa* digs a burrow to a depth of 26 65 cm, and sometimes seals the entrance with a clay plug. It chooses areas of heavy clay soil, in regions that quickly become flooded after rainfall. For the majority of the year, the crab remains in its burrow, even at night, although it may then come up to the burrow entrance. It emerges to feed only after heavy rains, and may therefore have to spend periods as long as a year without feeding. Its tolerance of water loss and of increases in the concentration of its body fluids have been discussed in Section 4.7.3. It survives by being able to take up water that condenses on its body while in the burrow: relative humidities at the bottom of the burrow remain between 80 and 100 % even during the day, and at night dew condenses as the temperature falls. The movement of the crab to the burrow entrance at night may aid this process by further cooling the body surface. Water that condenses on the carapace runs into the branchial chamber and is either absorbed there or passed to the mouth. During the dry season, *H. transversa* also reduces its metabolic rate and essentially enters a period of dormancy, so reducing its food requirements. In summary, *H. transversa* is truly a desert crab, paralleling the abilities of many other desert animals.

As discussed in Section 4.7.3, there are two possible ways in which the terrestrial line containing *H. transversa* could have evolved. If the invasion of land occurred via fresh waters, the physiological tolerance of *H. transversa* is hard to understand except as a modification developed *after* emergence from the water. The invasion of dry habitats would in this case provide a parallel with the vertebrates, but achieved by quite a different means. While the vertebrates have evolved a waterproof epidermis and a high degree of embryo protection, *H. transversa* still has a permeable cuticle, and needs to rear its young in fresh water. Its desert life is ensured only by behavioural and physiological adaptations linked to its fossorial habit. If, on the other hand, invasion of land occurred directly from the sea, and the relationship of the group with fresh water is only secondary, the high degree of tolerance is more easily understood. As pointed out in Section 4.7.3, it is possible that further work on the group may clarify this issue.

## 8.10     Conclusions

Freshwater animal lines have on several occasions given rise to terrestrial groups. In this section we can draw conclusions about two aspects of this evolutionary sequence: first, about the types of freshwater habitat most important in the evolution of terrestrial animals; and secondly, about the importance of freshwater habitats in this respect, when compared with marine habitats.

### 8.10.1     *Freshwater habitats and the invasion of land*

Reviewing some of the types of freshwater habitats at the beginning of this chapter, it was suggested that in comparison with lakes and swamps, rivers and streams were less likely to be sites where animals would evolve terrestrial habits. In the case of vertebrates and cyclophorid gastropods, it was then argued that adaptations to a swamp environment, with its reductions in dissolved oxygen, could have been pre-adaptive for life on land. It was pointed out, for instance, that adaptations for air breathing are common in swamp-dwelling animals, but rare in the inhabitants of fast streams. Considering other animal groups, however, it can be seen that such a view is a great over-simplification. Helicinid gastropods, for instance, are more likely to have been associated with hard surfaces, and particularly with the stems of emergent vegetation in well-aerated water, than with swamp habitats. Several other types of invertebrate have utilized the cryptozoic lifestyle to move out of fresh water. As well as the groups discussed here, some groups of small animals such as harpacticoid copepods and ostracods, and maybe others, have probably become terrestrial via the freshwater interstitial environment. Neither of these categories of animals would be likely inhabitants of swamps, because life within swamp sediments is usually very limited: not only is there no oxygen, but there is no access to surface (aerial) oxygen. The majority of groups such as ancestral earthworms, leeches and turbellarians, as well as the interstitial groups, probably moved on to land from lakes or slow-flowing rivers with a good oxygen supply.

The larger burrowers present a somewhat different case. Freshwater crabs and crayfish are not so dependent on the oxygen content of interstitial water as the smaller forms. They are to some extent found in swamps, particularly in regions where water cover is spasmodic. Little is known of the physical and chemical environment of the tropical freshwater crabs, but in temperate zones such as Tasmania, the water in crayfish burrows is often low in oxygen (Horwitz, Richardson & Boulton, 1985). This lowering of oxygen tension is not related to low oxygen in the water of the neighbouring creeks, but to a low rate of movement of water in and out of the burrows. The low oxygen levels in burrow water are not a hazard for the crayfish because most do not use this water for respiration: the water keeps respiratory membranes moist, but respiration takes place in the air-filled parts of the burrows. Decapod crustaceans may therefore have moved on to land from the edges of almost all freshwater bodies.

### 8.10.2 *Freshwater vs marine routes to land*

Comparing the animal groups discussed in this chapter with those covered in Chapters 5 to 7, it is readily apparent that far more terrestrial taxa have originated directly from marine ancestors than from freshwater lines. One possible reason for this has already been put forward at length: animals of marine origin have, in general, inherited physiological mechanisms of tolerance rather than regulation, while animals of freshwater origin have inherited regulatory mechanisms. Because of the very variable nature of conditions on land, tolerance mechanisms have, with some notable exceptions, been the more successful. Several other points may now be discussed.

First, there is a much wider phylogenetic base in the sea than in fresh water. As suggested earlier in this chapter, the estuarine environment can be regarded as a 'filter' that has prevented many groups from moving into fresh water. The picture is somewhat confused at the present day because the number of species in fresh water reflects not just those that have moved there from estuaries, but many that have re-invaded water from the land. Nevertheless, the summary presented in Table 1.1. still demonstrates that less than 50% of marine phyla have penetrated fresh water. This in itself would restrict the possibilities for the origin of terrestrial groups in fresh water. The origins and the composition of freshwater communities have been considered in detail by Gray (1988).

Secondly, it may be argued that the marine littoral zone presents a wider diversity of habitats than does fresh water. Most freshwater ecosystems, especially in lakes, swamps and the larger rivers, are dominated by fine sediments. The interstitial habitats of sand and regions of rock and shingle are relatively scarce and small in area. Where rock does occur, in fast streams, the fauna has to be a very specialized one, adapted for clinging to the substrate. The most common environments are areas of emergent vegetation with a soft substrate, and the origins of terrestrial lines may well have concentrated there.

Thirdly, the physical regimes in marine and freshwater shores are quite different. Most seas have regular tides, with part of the shore being uncovered usually once or twice a day. Marine littoral animals are therefore exposed to aerial conditions frequently, and on a regular basis. It is not surprising that many have become adapted to activity while the tide is out, especially those that live near the top of the shore, where submersion lasts for a relatively short time. On freshwater shores, in contrast, water level fluctuates on a seasonal pattern: this has often led to the adoption of long-term aestivation, but only in a few cases to short-term aerial breathing.

Lastly, there is the matter of time scale. Marine shores move slowly under the forces of erosion, deposition, uplift and submergence, but overall they stay in existence for long periods. Fresh waters and their shores, on the other hand, are much more transient phenomena. Lakes, in particular, have a shorter time span than rivers (Hynes, 1970). In the few present-day lakes that are of ancient origin, groups such as the gastropods have radiated widely (Boss, 1978), but in general freshwater bodies have short life-spans, and these allow little time for evolutionary processes to act.

# Part III

## Life on land

Although some consideration has been given to the ways in which the osmoregulatory properties of terrestrial animals have been dictated by their routes on to land (Chapter 4), there has been no overall approach to the physiological adaptations of animals on land. An attempt to remedy this is made in Chapter 9. Then in Chapter 10 the ecological characteristics of terrestrial communities are discussed, and comparisons are made with the communities of aquatic ecosystems.

# 9

## Terrestrial adaptations

Only an incomplete anatomical transformation from gill to air-breathing requiring frequent wettings in the salt ocean held them in place. Laboratory experiments have shown that they can survive with gills completely dissected out in dry air for several hours, an operation that would kill any normal sea-dwelling crab.

G. C. Klingel (1959) *Wonders of Inagua*, London: Robert Hale.

It is now essential to assess the adaptations of animals to life on land in a more detailed fashion than has been possible in previous chapters, and to compare the variety of adaptations found in different taxa. This will be done by taking various regulatory processes in turn.

### 9.1  Osmoregulation and water balance

The use of two fundamentally different approaches – tolerance and regulation – which allow animals to survive changes in the water and salt content of the immediate environment have already been described. Despite the differences between the two approaches, it is evident that even those animals depending essentially upon tolerance mechanisms must also have available *some* regulatory mechanisms. In other words, there must in all animals be a balance between water and salt gained, and water and salt lost. The physiological, structural and behavioural mechanisms that bring about such balance can be considered under a number of headings. It is impossible to deal with *all* animal groups, and here a selection will be made.

### 9.1.1  Salt and water uptake

Aquatic animals have, by definition, an immediate supply of water and salts, although depending upon the animal's osmotic relationship to the medium, these may be more or less readily available (Section 2.2). In the sea, most invertebrates are slightly hyper-osmotic to the outside medium. They thus generate an inward flow of water that maintains their body turgor (see, e.g. Little, 1981). In fresh water, all animals are hyper-osmotic to the medium, and again water flows into their bodies osmotically. A supply of salts is usually achieved by active uptake processes, often located in the gills: these directly absorb ions from the water.

On land, the supply of salts and water is usually erratic, and most animals probably have specific mechanisms for obtaining both. Unfortunately this subject has been ignored in many animal groups, and preference has been given to measuring rates of water loss, a much easier task. In many terrestrial invertebrates, the food is suspected to be the major source of both ions and water. Particularly in such environments as the leaf litter of tropical rainforests, decaying leaf material ensures a fairly constant supply of both.

The specific adaptations so far investigated in terrestrial and semi-terrestrial animals all involve the uptake of water and not the uptake of salts. Among the crustaceans, many crabs can absorb water through the gills, simply by immersing themselves in pools. When only damp sand is available, several species of grapsids, ocypodids and gecarcinids can squat on the sand surface and absorb water into the gill chambers by the capillary action of hairs on the ventral surface of the body or on the legs (Powers & Bliss, 1983). The shore-living isopod *Ligia oceanica* can also use capillary action (Hoese, 1981), and terrestrial woodlice can absorb water from porous surfaces both through the mouth and the anus (Spencer & Edney, 1954). The most extreme absorption mechanism has been reported for the crab *Ocypode* (Wolcott, 1976). By muscular expansion of the gill chamber, pressure inside it is reduced by as much as 40 mm Hg, and moisture is literally sucked from the sand. A parallel to this is found in some spiders (Parry, 1954). *Alopecosa accentuata*, when it is desiccated, can drink against a suction pressure as high as 350 mm Hg (Fig. 9.1).

A second type of adaptation to absorb water is provided by the so-called coxal sacs of many species in the insect – myriapod – onychophora assemblage. These are best understood in some of the apterygote insects such as the thysanuran *Petrobius brevistylis* (Houlihan, 1976). This animal lives in rock crevices in the marine supralittoral zone. It has 22 coxal sacs, which are paired abdominal organs, capable of being everted and placed on the damp substrate. An active process, probably initiated by the movement of salts, moves water into the sacs. Similar sacs are found in other apterygotes, some myriapods and in the onychophorans.

The third way in which water can be taken up is as a vapour. Some of the woodlice can probably take up water from saturated air, and desert beetles can collect dew as it falls. *Onymacris unguicularis* lives in the Namib Desert

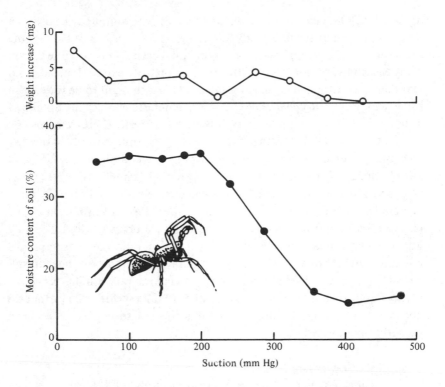

Fig. 9.1. The relation between moisture-suction characteristics of soil and the ability of a spider, *Alopecosa accentuata*, to take up water. The lower graph shows moisture content of an artificial soil at various suction pressures. The upper graph shows mean water uptake by spiders. Spiders can suck water out of the soil even when the moisture content is as low as 20%, and against pressures of over 300 mm Hg. After Parry (1954).

of southern Africa. When there are dense fogs, this beetle climbs to the crest of sand dunes, and stands with its head down and abdomen up, so that water condensing on the body runs down towards the head and can be drunk (Hamilton & Seely, 1976). Even more intriguing is the ability of some insects and arachnids to take up water from air that is not saturated with water vapour. The apterygote *Thermobia domestica*, the firebrat, can remove water from atmospheres well below saturation. Water vapour is absorbed by the lining of the rectum. Several pterygote insects such as the bird lice, book lice, fleas and some beetle and cockroach larvae can also absorb water vapour, but strangely this ability is limited to wingless stages (Edney, 1977). Apart from the larva of the beetle *Tenebrio molitor*, the rectum is not involved. In the desert cockroach *Arenivaga investigata*

(O'Donnell, 1982) two eversible bladders below the mouth may be used. In others, uptake may occur directly across the general cuticle. Whatever the site, however, the general mechanisms are not certain. It is possible that in many cases a hygroscopic material is secreted on to an external surface, and that the water attracted is then swallowed. This is thought to be the case in some ticks, and in *Bryobia praetiosa* this mechanism enables the uptake of water vapour from humidities as low as 50% R.H. There remains the problem of how water is removed from the hygroscopic material in the gut.

A final source of water is the oxidative metabolism of fats and carbohydrates. The classic study here is that of Schmidt-Nielsen (1964) who investigated some small desert rodents. He showed that in most mammals, extra oxidation processes necessitated extra ventilation of the lungs, and consequently incurred increased respiratory water loss. The only way in which animals could achieve an overall gain in water was to reduce the water content of expired air using a nasal countercurrent mechanism, and such mechanisms are present in desert kangaroo rats. Similar processes may occur in such insects as locusts (Little, 1983), but for animals that do not ventilate their respiratory apparatus the situation has never been properly investigated.

### 9.1.2   Water-proofing

Table 9.1 gives some comparative figures for the rate at which water is lost across the body surface for some groups where rates have been measured. The values are all given as integument permeability, i.e. the loss of water per unit time per unit saturation deficit (see 2.4.3). While this measurement has some drawbacks, as will be discussed later, it allows some initial comparison of the properties of the integument in a variety of animals.

The highest rates of water loss are shown in the snail *Helix aspersa*, the frog *Rana pipiens* and the urodele *Desmognathus ochrophoeus*. Few measurements are available for other 'soft-skinned' animals, but rates are probably similar to those of snails and amphibians because these are similar to those expected from a free water surface. Such high rates are seldom maintained for long, however, for two major reasons. First, evaporation depends upon the maintenance of a large gradient in water concentration between air and evaporating surface, and the evaporating water tends to reduce this gradient. The gradient is also reduced because of a thin 'boundary layer' next to the skin, where the air is unstirred and humidity tends to remain high. The second point concerns factors *inside* the animal. Maximum evaporation rates depend upon an instantaneous

Table 9.1 *Integument permeability in some terrestrial and semi-terrestrial animals*

| | Permeability $\mu$g.cm$^{-2}$.h$^{-1}$.mmHg$^{-1}$ |
|---|---|
| ***Mollusca*** | |
| Pulmonate gastropods | |
|   *Helix aspersa* | 2500 |
|   *Otala lactea* (mantle collar, during aestivation) | 16 |
| ***Crustacea*** | |
| Decapods | |
|   *Uca marionis* | 200 |
|   *Uca annulipes* | 80 |
| Isopods | |
|   *Ligia oceanica* | 220 |
|   *Venezillo arizonicus* | 15 |
| ***Myriapoda*** | |
| Diplopods | |
|   *Glomeris marginata* | 200 |
|   *Orthoporus ornatus* | 8 |
| ***Arachnida*** | |
| Ticks | |
|   *Ixodes ricinus* | 60 |
|   *Ornithodoros delanoei acinus* | 0.4 |
| Scorpions | |
|   *Pandinus imperator* | 76 |
|   *Hadrurus arizonensis* | 1.2 |
| ***Insecta*** | |
| Apterygotes – collembolans | |
|   *Tomocerus problematicus* | 701 |
|   *Seira domestica* | 3.7 |
| Apterygotes – thysanurans | |
|   *Thermobia domestica* | 15 |
|   *Ctenolepisma terebrans* | 0.7 |
| Pterygotes – lepidopterans | |
|   *Hepialus* larvae | 190 |
| Pterygotes – beetles | |
|   *Tenebrio molitor* larvae | 5 |
|   *Tenebrio molitor* pupae | 1 |
| Pterygotes – dipterans | |
|   *Glossina morsitans* adults | 8 |
|   *Glossina morsitans* pupae | 0.3 |
| ***Vertebrata*** | |
| Amphibians | |
|   *Rana pipiens* | 2410 |
|   *Desmognathus ochrophoeus* | 4120 |
| Reptiles | |
|   *Caiman sclerops* | 70 |
|   *Xantusia vigilis* | 2 |

*Note:* Maximum actual rates of loss in air are shown by pulmonate gastropods and amphibians, and these approach the theoretical maximum loss rate from a free water surface.

*Source:* Edney, 1968, 1977; Little, 1983; Machin, 1975, 1977.

replacement of water at the integument surface from deeper tissues. There is usually some delay in replacing surface water, so that evaporation rates slow down with time. Another major mechanism relating to water loss in soft-bodied animals is the production of mucus. This can be rapidly secreted on to the body surface by glands that draw their water supply from below the epidermal layers. In pulmonate snails, for instance, this allows the epidermis to be continuously lubricated from outside, promoting high rates of evaporation (Machin, 1977), and a similar mechanism probably occurs in animals such as flatworms, nemertines, oligochaetes and amphibians.

From what has been said above, it is not surprising that evaporation rates from soft-skinned animals should decrease as the integument dries up and as external air-mixing processes decrease. It *is*, however, extraordinary that rates as low as that shown for the snail *Otala lactea* should occur during aestivation: rates 100 times lower than those in normal, hydrated, mucus-covered snails. This ability is restricted to the mantle collar of aestivating terrestrial pulmonates – the part of the snail nearest to the shell aperture. The mechanism involved remains obscure, but appears to be due to some property of the cells of the mantle epithelium (Appleton *et al.*, 1979) which actively reduces water loss. Similar low rates of water loss are found in *Helix* (Fig. 9.2).

In comparison with the soft-skinned animals, those with a hard cuticle have a generally lower integument permeability (Table 9.1). Within each taxonomic group there is a great range of values, but even the highest of these is usually no more than one tenth of those in soft-skinned animals. This reduction in permeability is due to cuticular structure, which physically retards diffusion of water from the tissues to the atmosphere. Many decapod and isopod crustaceans and apterygote and pterygote insects show a considerable reduction beyond this, but interest has, understandably, centred on those species that have reduced permeability by another 1 or 2 orders of magnitude. Examples of these can be found in most taxonomic groups of animals with cuticles, and in particular in species living in deserts or other xeric conditions. They include the desert isopods like *Venezillo arizonicus*, myriapods like *Orthoporus ornatus*, ticks and scorpions, and both pterygote and apterygote insects. The lowest permeability has been found in the pupae of the tsetse fly *Glossina morsitans*, with a value of 0.01% of that shown by hydrated snails. This waterproofing appears in most cases to be due to an epicuticular wax layer: a very thin layer of wax, perhaps only one molecule thick, that is secreted on the surface of the cuticle. The mechanism of action of this highly efficient

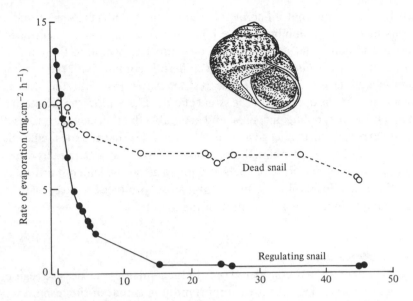

Fig. 9.2. The rate of evaporation of water vapour from an inactive snail, *Helix aspersa*. After initial high rates of loss, the living individual lost water at less than 10% of the rate found in a dead snail. This reduction in loss is due to a mechanism present in the mantle collar. After Machin (1966).

system has been discussed in detail by Edney (1977), and a broad survey has been given by Lillywhite & Maderson (1988). Here it is perhaps more important to discuss how such a mechanism might have arisen.

One possibility derives from work on the apterygote insects. In these it has been suggested that the epicuticular waxes arose under selection pressures acting on primitive forms within humid cryptozoic habitats. Although desiccation was probably not a major problem in these conditions, waterlogging could have been a major selective force: saturation of the leaf litter habitat, for instance, would have led to drowning of many types of invertebrate. In some present-day collembolans, the body surface bears many tubercles, each covered with an external wax layer. These tubercles effectively render the body hydrofuge, so that it maintains an air layer round it even under water, allowing gas exchange to continue. It is possible that such hydrofuge areas provided the basis for the adoption of a wax layer over the whole body in later evolutionary lines which left the cryptozoic environment, when desiccation became a dominant selective pressure.

Waterproofing is also a characteristic of some of the terrestrial vertebrates (see Lillywhite & Maderson, 1988), but the mechanisms involved are somewhat different from those of the insects. Semi-aquatic forms such as the amphibians have high rates of water loss, as mentioned above, and numerous glands that produce mucus open on to the skin. In the mammals, the outer layers of the epidermis contain both the protein keratin and layers of lipid. These restrict water loss, but the skin is punctuated by the openings of the sweat glands. These allow the facultative loss of water for cooling purposes, and inevitably the skin can never form a good barrier against water loss. It is in the reptiles that the most efficient mechanisms of waterproofing have evolved. Their major barrier to water loss lies in the inner layer of the stratum corneum, which contains both keratin and phospholipid. This combination produces a layer with a permeability almost as low as that of the insects.

### 9.1.3    *Respiratory water loss*

So far we have considered only the general body surface as a region of potential water loss. Another important area is that of the respiratory organs. These are kept moist so that gases can diffuse across the respiratory membranes, and water inevitably evaporates from them. The general evolutionary trend of terrestrial animals has been to replace external gas-exchangers with internal ones: the gills of aquatic animals are replaced by lungs. Water loss from lungs is often still high, however, because each time the lung is ventilated moist air is breathed out. This is a particularly obvious problem in animals such as the vertebrates with their complete dependence upon respiratory ventilation. In the insects, which have a tracheal respiratory system, diffusion plays a larger role, and respiratory water loss can be reduced by closing the tracheae. Many insects can close the spiracles – the places where tracheae open on the body surface – and in some pupae the spiracles stay closed except for brief periods. In spiders, too, the entrance to the lung-books can be at least partly blocked. Myriapods, on the other hand, have no true mechanism for closing the spiracles, and this is probably an important factor in reducing the degree to which they have invaded very dry environments.

It is instructive to have an example in which the importance of respiratory water loss can be seen. In the desert tenebrionid beetle *Eleodes armata*, respiratory losses at 25 °C accounted for only 3% of total evaporative water loss (Ahearn, 1970). At 35 °C this rose to 21%; and at 40 °C loss through the spiracles was 40% of the total. In some circumstances, then, even the efficient mechanisms of water retention found in the insects begin to become inefficient.

### 9.1.4    Renal water loss

All animals need to excrete nitrogenous waste, and to have a mechanism for getting rid of excess water from the body. In most cases these processes are carried out in the excretory organs or kidneys. These organs are therefore responsible for some loss of water. Terrestrial animals show a great variety of ways in which this water lost through the kidneys is kept to a minimum. Here we will not attempt a complete survey, but will take five examples. As will be seen later (9.1.5), many excretory organs work in conjunction with the gut, and water absorption occurs in the rectum. This section will deal solely with salt and water retention by the excretory organs themselves.

The first example is taken from some of the animals typical of the cryptozoic fauna. In the flatworms and nemertines, the excretory tubule commences as a flame cell, with a bundle of partially fused flagella beating within a chamber which has walls typical of an ultrafiltration site. Urine passes down the tubule, and the walls of this are just like other tubules known to be involved in re-absorption. The overall organization of the flame cell systems in terrestrial flatworms and nemertines seems to provide a system whereby excess water entering the body can be rapidly excreted, and the importance of the re-absorptive structures is unknown. It seems likely that the systems excrete hypo-osmotic urine, and are adapted not to conserve water, but to excrete it as fast as possible. The protonephridial systems seem very appropriate for animals living a very nearly freshwater existence, with a cryptozoic lifestyle in a habitat where too much water is more likely to be a problem than too little.

The second example is provided by a comparison between pulmonate and prosobranch gastropod molluscs. The terrestrial pulmonates show a wide variation in renal structure, and probably had several separate lines of evolution (Delhaye & Bouillon, 1972). The majority of present-day species are placed in the suborder Sigmurethra, characterized by a kidney opening into an elongate ureter which is reflected to form a primary and a secondary arm. The secondary arm opens near to the pneumostome, outside the mantle cavity. The terrestrial prosobranchs such as the Cyclophoridae have no ureter, and the kidney opens at the back of the mantle cavity. In the pulmonates, where details have been worked out for *Helix* and *Archachatina* (Vorwohl, 1961), urine is initially formed by filtration directly into the kidney. The cells lining the kidney lumen also excrete uric acid concretions which are carried into the ureter. Salts and water can then be re-absorbed in the ureter: in wet surroundings a copious flow of hypo-osmotic urine is produced, while in the dry a 'dry urine' of uric acid is extruded from the renal pore. The snails are not able to produce hyper-osmotic urine. The spatial separation of uric acid production and water re-absorption is an important adaptation to life in surroundings of variable climate: even in dry situations, the kidney contents are flushed into the ureter, and waste nitrogen can be taken directly to the outside.

Terrestrial prosobranchs such as the cyclophorids show no such separation between the sites of various excretory functions. Urine is produced not in the kidney but in the pericardium, and it then drains into the kidney lumen. Its production cannot be controlled as carefully as in the pulmonates, since it is intimately linked with circulatory processes. Water is re-absorbed in the kidney, in the same site as the production of uric acid. There is no ureter, and it is therefore not possible for cyclophorids to excrete uric acid without losing a great deal of water. These differences between pulmonates and prosobranchs probably came about in response to selection pressures during early evolution. For pulmonates, with a marine saltmarsh ancestry, there would have been a high priority for the ability to excrete nitrogen even in a dry environment. For prosobranchs such as the cyclophorids, with a freshwater origin, the highest priority was to produce a high flow-rate of hypo-osmotic urine. This flushed uric acid out of the kidney, and there was little selection for the spatial separation of nitrogen excretion and water balance.

The third example of the importance of renal organs in salt and water balance concerns the semi-terrestrial crabs. Since many semi-terrestrial decapods can maintain the osmotic pressure of their blood well below that of the external media, it might be expected that they could produce hyper-osmotic urine. This is not the case, however, in any known crab species. The antennal glands are important in regulating ionic composition of the blood, especially the concentrations of magnesium and sulphate, but urine is always isosmotic with the blood. Salt is probably excreted extrarenally, either through the gills or through specialized organs of the gut. The antennal glands can alter the rate of excretion of urine, so that when *Ocypode albicans* is placed on dry sand, urine production stops (Flemister, 1958). This is not, however, due to re-absorption in the renal system, but to the cessation of the initial ultrafiltration process. In this case, then, the excretory organs appear unable to exert control over water re-absorption, and the whole system must be turned off to prevent loss of water.

The excretory organs of many terrestrial arthropods bear a striking similarity to the antennal glands of crabs. In myriapods and apterygote insects, the maxillary glands open under the head. In onychophorans there are serial excretory organs that are modified for saliva production in the head, but are used for filtration and re-absorption in the more posterior segments (Gabe, 1957). Arachnids have organs called coxal glands, opening at the base of the walking legs. Although these are not homologous with the maxillary glands of insects and myriapods, they have very similar characteristics. Only in the pterygote insects are such excretory organs lacking (see Section 9.1.5).

The maxillary glands consist of two sections. One, known as the sacculus, is the region in which urine is produced by ultrafiltration through podocytes – cells with thin membranous areas similar to those in the vertebrate glomerulus and in most invertebrate excretory organs. The second section is a long tubule with varying structure, in which salts and water are thought to be re-absorbed. It may seem surprising that most evidence about the function of these organs comes from fine structure, but the glands are usually very small, and not easy to investigate physiologically. In the coxal glands of a tick, however, Kaufman *et al.* (1982) have demonstrated ultrafiltration into the first section of the gland, the 'filtration sac'. While the force producing filtration in this case is produced by muscles that stretch the sac from the outside, lowering the pressure inside, the mechanism of filtration in maxillary glands is unknown. Evidence for re-absorption in the tubule is confined so far to fine-structural similarities with known re-absorbing systems.

The adaptations of the renal systems of vertebrates have been discussed in many text books, and there is no need to deal with them in detail here, but it is interesting to follow some of the general trends from fish to amphibians and then to reptiles, birds and mammals. Freshwater fish of all taxonomic groups produce hypo-osmotic urine from the kidneys, and so remove from their bodies water that enters osmotically. At the same time, they compensate for the loss of ions by actively absorbing salts through their gills. The early amphibians probably used very similar mechanisms. Present-day forms also produce hypo-osmotic urine, but at least in frogs, which have no gills, the site of salt absorption is now in the skin. The system in reptiles is rather different. Because reptiles have an epidermis that is relatively impermeable to water, the need to excrete water is reduced, and the major requirement is for water conservation. This is aided by the production of isosmotic urine, which allows the excretion of waste metabolites without great water loss. A further problem in very dry climates is the removal of excess salts, but reptiles are unable to produce hyper-osmotic urine. They excrete salts extrarenally, through nasal glands (Peaker & Linzell, 1975). Most birds and mammals can produce hypo-osmotic or isosmotic urine, but like the reptiles cannot concentrate the urine. Some of the most interesting exceptions to this rule are the desert mammals which have specialized kidneys. Instead of the normal kidney tubules, each tubule has a 'loop of Henle' interpolated between the proximal and distal convoluted tubules. In this a countercurrent concentrating system produces fluids several times more concentrated than the blood. By controlling the permeability of the collecting ducts, the final urine may be allowed to equilibrate with these high concentrations, and hyper-osmotic urine is excreted. The kangaroo rat, *Dipodomys*, for instance, examined in detail by Schmidt-Nielsen (1964), can produce urine 14 times more concentrated than the blood. It therefore loses very little water even when excreting large quantities of nitrogen.

### 9.1.5    *Water loss from the gut*

Water contained in the faeces provides another possible route of water loss for all animals. Only in a few groups has this subject been studied intensively, but the general involvement of the gut in water balance can be considered here in several examples, particularly those in which the functions of the excretory organs and rectum are in some way combined.

The oligochaetes provide an initial example of this combination of excretory with digestive processes. Earthworms possess segmental excretory organs, the nephridia, which usually open directly to the outside. They are therefore a potential source of water loss. In some of the earthworms found in drier environments, such as *Pheretima*, the nephridia open instead into the intestine (Bahl, 1947). The faeces of these earthworms are much drier than those of other species, presumably because urinary water is re-absorbed from the gut. The whole process needs further study, but seems to be an adaptation to dry soils.

In woodlice, the function of the gut in water balance is rather more complex. Most woodlice live in damp environments so that restriction of water loss is not often critical for them. They respire through external gills formed from the pleopods, and these are kept moist by urine from the maxillary glands, distributed over the under surface of the body by a series of capillary channels. This water is not, however, lost to the animals because it can be re-absorbed by being taken into the rectum. Water is therefore effectively circulated from maxillary glands to rectum – a parallel to the situation in earthworms, except that here the water circulates externally. In some desert woodlice, where water retention is critical, the rectum is modified for water absorption, and *Hemilepistus* spp. can absorb significant quantities of water by eating damp sand (Coenen-Stass, 1981).

By far the most attention has been paid to the importance of the rectum of insects in water absorption. The excretory organs of pterygote insects are the Malpighian tubules, which open into the gut, and, while these provide a flow of urine, it is the rectum that is responsible for re-absorbing water. Similar systems are found in many apterygote insects, in addition to the maxillary glands which open directly to the outside, and in the myriapods, but so much more is known about the pterygote system that discussion will here be restricted to this.

The fluid produced by the Malpighian tubules is not identical with haemolymph, but is very similar. Thus, although it is slightly hypo-osmotic and has a higher potassium content, the major influence upon what is finally excreted is exerted by the re-absorption processes in the rectum. This re-absorption is brought about by a variety of mechanisms in different insect types. Most insects, for example the cockroach *Periplaneta americana* and the blowfly *Calliphora erythrocephala*, have rectal pads or papillae: specialized areas of the rectal wall in which spaces between the epithelial cells provide a compartment separate from both the rectal lumen and the haemolymph. Ions or other molecules are transported into this space from the rectal lumen, and water then follows down the osmotic gradient (Wall, 1977; Berridge & Gupta, 1967). The fluid accumulating in the intercellular spaces is then forced into the haemolymph. In locusts this process of water re-absorption can reduce the water content of the faeces to as little as 39%. Fig. 9.3 shows that water can be absorbed by the rectum of *Schistocerca* across a wide range of osmotic gradients.

In some Coleoptera and in larval Lepidoptera, the mechanism is more complex and not so well understood. There are no rectal pads, but the outer surface of the rectum is invested with the blind ends of specialized Malpighian tubules, which are further covered by a membrane. These tubules are called cryptonephric tubules, and unlike the normal tubules they can modify the ratio of sodium:potassium in their fluid. This may allow insects to excrete large quantities of ions taken up in the food. In addition to direct effects on ionic composition, some ions are moved from the tubules into the space adjoining them, covered by the membrane. The fluid in this space, the perinephric space, is increased in osmotic pressure. As in the situation with the rectal pads, water moves down the osmotic gradient from the rectal lumen into this intermediate compartment, and thence into the haemolymph. The larvae of the beetle *Tenebrio molitor*, using this mechanism, can withdraw water from the faeces until equilibrium is reached with 75% R.H. (Ramsay, 1964) – representing an osmotic pressure more than five times that of sea water. The exact details of this

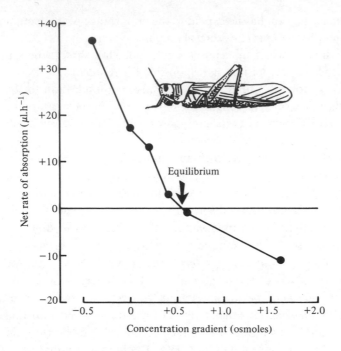

Fig. 9.3. Absorption of water by the rectum of the desert locust, *Schistocerca gregaria*. The graph shows rate of absorption in an isolated preparation, plotted against the osmotic gradient across the rectal wall. Water can be absorbed even when there is an osmotic gradient of more than 0.5 Osm, and this process dries the faeces before they are ejected. After Phillips (1964).

mechanism are still obscure, however, and remain a challenge for further work.

Water can probably be absorbed by the gut in all vertebrates, from fish to mammals, but the importance of this absorption, and its site, vary widely between the classes. In marine teleost fish, for example, water is taken up through the intestine to offset water lost osmotically to the outside. In amphibians, the gut may also take up large quantities of water, but in terms of water conservation, the gut is most important in reptiles. Here the cloaca is of major importance in absorbing water from both faeces and urine. Such post-urinary absorption ensures that, in desert forms, very little water is lost (Minnich, 1979). Since nitrogen is excreted in the form of urate salts in reptiles, a high proportion of the potassium that is excreted is precipitated in the cloaca as solid potassium urate when the water is re-absorbed.

### 9.1.6  *Behavioural adaptations, and overall water balance*

General behavioural adaptations to life on land are considered in Section 9.7, but here we can discuss those especially concerned with water balance. The crypotozoic fauna are characterized by their ability to remain in humid niches, thus conserving body water. Terrestrial flatworms and nemertines, for instance, as well as many woodlice, have behavioural reactions that take them to areas of high humidity and darkness, and where the body has intimate contact with the substrate. Very often the activity of these animals peaks at night time, and this activity is regulated by endogenous rhythms. For some of the smaller animals such an approach is not sufficient, and many nematodes, tardigrades, rotifers, some branchiopod crustaceans, and some insect larvae, can survive being dried in the state termed cryptobiosis. In this state, they can tolerate extreme temperatures, and many can survive for long periods. The cysts of the brine shrimp, *Artemia salina*, for example, can tolerate 780 days at $0°C$ (Sundnes & Valen, 1969), and have been shown to survive in the dried state for 28 years. After this, they rapidly re-hydrate and begin development when placed in water. The chironomid larva *Polypedilum vanderplanki*, found in temporary pools in Africa, can withstand temperatures from $-270°C$ to over $+100°C$ (Hinton, 1968). Although the mechanism of cryptobiosis can not be said to be well understood, it allows a variety of animals to survive quite extreme terrestrial conditions.

Most terrestrial animals, however, merely become inactive when conditions become unsuitable. The hygrophilic fauna, such as the gastropod molluscs, emerge when the substrate is damp, often cued by a fall in temperature (Runham & Hunter, 1970). Their inactive state is known as aestivation (in summer) and hibernation (in winter). Some snails can survive even desert climates by entering this inactive phase, and the limit to the period of aestivation then seems to be set not by physical extremes, but by lack of sufficient food supply to continue basal metabolism (Schmidt-Nielsen *et al.*, 1971). As with cryptobiotic animals, aestivating species usually emerge rapidly from the inactive state when conditions improve.

Some quite sophisticated behavioural mechanisms of osmoregulation have been described in the decapod crustaceans. The hermit crab *Coenobita perlatus*, for example, maintains a water supply in its host shell. At night time it moves down to the sea to renew this supply, and it can regulate its own haemolymph composition by choosing fresh and salt water in appropriate proportions (Gross & Holland, 1960). The true crab *Pachygrapsus crassipes* does not carry a large supply of water with it, but, by

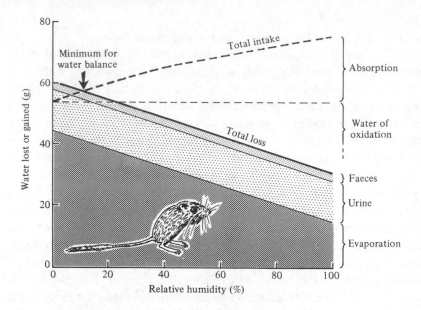

Fig. 9.4. Overall water balance in a kangaroo rat, *Dipodomys* sp., over a range of humidities. Water gained from oxidation is independent of humidity, but at higher humidities food contains more water absorbed from the atmosphere in the burrow. Water losses in faeces, urine and by evaporation are related linearly to humidity. The totals of loss and gain cannot be balanced below a humidity of about 10%, but above this the animal can remain in water balance without drinking. After Schmidt-Nielsen (1964).

spending variable amounts of time in fresh and salt water, it too can absorb the right proportions to maintain a constant haemolymph composition (Gross, 1957).

Overall water balance is therefore affected by a large number of factors, but in the end may be regulated behaviourally. To gain an idea of how overall regulation may be achieved, we can end this section by considering a desert mammal, the kangaroo rat *Dipodomys* (Schmidt-Nielsen, 1964). In the desert environment, kangaroo rats lose water mainly by evaporation from the skin and the lungs, although some potential respiratory water loss is avoided by a countercurrent heat exchanger in the nasal passages which cools expired air and therefore reduces its water content. Losses in faeces and urine make up the total loss, which has to be balanced by water of oxidation gained from the metabolism of food, and water absorbed in the food itself – kangaroo rats seldom actually drink. Water of oxidation is particularly valuable because, with the nasal countercurrent mechanism, respiratory losses are minimized. Absorbed water is also important, however, and is increased because the kangaroo rat stores food in its burrow, where relative humidity is high: the food itself therefore absorbs water from the air before being eaten. With this mixture of physiological and behavioural adaptations, summarized in Fig. 9.4, kangaroo rats can live in extremely desiccating environments.

## 9.2    Respiration

It has been argued in Chapter 8 that the development of air breathing in fresh water probably began in swampy regions with low oxygen tensions. Such situations are uncommon in the present-day marine intertidal zone, although some authors believe that in the past the lack of oxygen in marine swamps may have been important in the origins of air breathing (Packard, 1974). Nevertheless, air breathers are common on present-day marine shores, where selection pressures are quite different. Presumably it has been not the lack of dissolved oxygen, but factors such as the availability of new food sources and the possibility of escape from predators that have led to the development of air breathing in these regions that are covered and uncovered rhythmically by the tides. Many modern intertidal gastropods, for instance, feed at the time of low tide, thus avoiding predatory marine fish and crabs (Little 1989a). Their responses to periodic emersion, in terms of respiration, vary with taxonomic group (McMahon, 1988). The behavioural, morphological and physiological adaptations of intertidal fish have been reviewed by Bridges (1988). The selection pressures acting on high-shore crabs have been discussed by Wolcott & Wolcott (1985). Competition for food and burrow sites, intraspecific aggression, and predation by vertebrates are all possible factors, but it must be admitted that we do not really understand *why* crabs of marine origin migrate inland.

This section will consider first the differences between air-breathing systems evolved in freshwater and marine environments, and will then describe some of the differences between systems adapted for air breathing and water breathing.

### 9.2.1    *Air breathing in animals of marine and freshwater origin*

The gas exchange systems evolved in marine intertidal situations seem in broad terms to be similar to those found in animal lines derived from freshwater ancestors. Aquatic respiratory systems in both fresh and salt water are usually based on external exchange sites – the gills – while animals out of water respire through internal exchange sites – the 'lungs'.

This has probably come about for two reasons. First, gill filaments tend to collapse in air, so that their surface area is much reduced, and their exchange efficiency declines. Secondly, an external exchanger is very vulnerable to desiccation, providing a large area for evaporative water loss. The development of an internal cavity counters both these problems to some extent.

Nevertheless, there is one common difference between the lungs of animal lines of freshwater origin and those of direct marine origin, related to the availability of air for breathing. In stagnant fresh water, animals must come to the water surface to breathe, yet for most of their life functions, such as feeding, reproduction and dispersal, they must stay below the surface. They must, therefore, by definition, be *intermittent* air breathers. Details of some of these intermittent mechanisms will be given later. In contrast, intertidal marine animals spend many hours out of water and must wait for tidal submersion to respire aquatically unless they move back into rockpools or burrows. They tend, therefore, to be *fairly continuous* air breathers. This division is by no means a rule, since fairly continuous air breathing (e.g. buccal respiration in the vertebrates ) *has* evolved in fresh water; but in contrast intermittent air breathing is very uncommon in the marine intertidal zone. The consequences of this division will be discussed later.

### 9.2.2   The variety of mechanisms of oxygen uptake

Although most aquatic animals take up oxygen through gills, many also take up a significant proportion across the general body surface. Indeed, many of the smaller invertebrates such as oligochaetes and platyhelminthes have no differentiated gas-exchange organs: with a large surface area:volume ratio, sufficient oxygen is obtained by diffusion through the epidermis. Larger soft-skinned invertebrates such as the molluscs also obtain some oxygen through the skin: in the pulmonate gastropod *Lymnaea stagnalis*, cutaneous oxygen uptake accounts for about 25% of the total uptake at low external oxygen tensions, but, as external oxygen rises, the skin can be responsible for more than 50% uptake (Jones, 1961). The use of the skin in oxygen uptake also occurs in air. Many intertidal prosobranch gastropods, for example, have well-vascularized parts of the mantle skirt and the floor of the mantle cavity, so that when air is drawn into the mantle cavity, oxygen is absorbed more efficiently than it would be through the collapsed gills.

The importance of the skin in oxygen uptake can be seen in the vertebrates by considering the common eel, *Anguilla anguilla*. This species can survive in air for weeks if the temperature is below 7°C. Its rate of oxygen consumption is halved compared with that in water, and 60–70% of oxygen is absorbed through the skin. This is achieved by accentuating blood flow through the skin capillaries. It is important to notice that, as temperature rises, the skin becomes unable to supply sufficient oxygen, and more efficient systems of obtaining oxygen are required.

In general, terrestrial animals possess some kind of internal gas exchanger. In the majority of invertebrates, and in the vertebrates, this is called a lung, although it has arisen independently many times and is therefore not a homologous structure except within groups. The gastropod molluscs provide good examples of the evolution of a lung. Some, such as the amphibious prosobranch *Pomacea*, have a divided mantle cavity with a gill in one half, and the other half acting as a lung. This animal can breathe air and water, simultaneously. When in de-oxygenated water, it forms a siphon with a flap of mantle tissue, protrudes this through the water surface, and then pumps air into the lung by rhythmic movements of the body (McClary, 1964). It therefore possesses what is known as a ventilation lung, in contrast to the lungs of many truly terrestrial prosobranchs, which are thought to be diffusion lungs. In these, the lung is not actively ventilated, and oxygen is gained solely by diffusion. Unlike these prosobranchs, the terrestrial pulmonates such as *Helix aspersa* have true ventilation lungs, drawing in air by alternately lowering and raising the floor of the mantle cavity (Ghiretti & Ghiretti-Magaldi, 1975). Some of the pulmonate slugs, in the family Athoracophoridae, have quite complicated lungs: the lung cavity is no longer a simple invagination lined with blood vessels, but possesses fine diverticula which radiate into a blood sinus, and provide a large area for gas exchange.

The terrestrial isopod crustaceans provide an exception to the idea that gills need to be replaced by lungs on land (Sutton & Holdich, 1984). Aquatic isopods obtain their oxygen through the pleopods – thin, platelike appendages on the underside of the abdomen that beat regularly to maintain water flow over themselves. The high-shore littoral form *Ligia oceanica* obtains about 50% of its oxygen through the pleopods, and the rest through the ventral surface of the abdomen (Edney & Spencer, 1955). Terrestrial woodlice also breathe through the pleopods, and *Porcellio scaber* obtains as much as 70% of its oxygen through them. Rates of oxygen uptake differ markedly in moist and dry air, and here again it appears that internalization of the gas exchange surfaces is involved. Many of the terrestrial woodlice have 'pseudotracheae' inside their pleopods – branched cavities opening to the outside by a single pore. Species possessing these organs are much more efficient at extracting oxygen from dry air than those without, although it is not known exactly how this increased efficiency is achieved.

The larger Crustacea do not depend upon external pleopods, but obtain oxygen through internal 'lungs' – usually well vascularized parts of the chamber housing the gills (McMahon & Wilkens, 1983). Crabs that are not specialized for air breathing, such as the intertidal *Carcinus maenas*, can ventilate the gill chambers in air as well as in water, using the beating of an appendage called the scaphognathite. The relatively thick gill lamellae are not very permeable, but at least do not collapse easily in air. More specialized crabs have mechanisms for aerating the water that remains in the gill chamber. The grapsid *Cyclograpsus punctatus*, for example, expels water from the gill chambers through the exhalant openings. This water runs in a thin film over the frontal parts of the carapace, held in place by specialized geniculate hairs, and then trickles back into the gill chamber through openings at the bases of the legs. Alternatively, air can be pumped through the water retained in the gill chamber, as in some ocypodid crabs.

The gills of crabs are progressively reduced in air-breathing species, and the dependence on the lungs becomes more acute. In relation to the type of gas exchanger, the oxygen affinity of the blood changes markedly. As air flow becomes predominant, and more efficient, haemocyanin aids oxygen absorption and high arterial oxygen tensions can be reached (Fig. 9.5). The details of the evolution of gas exchange organs have been discussed by Innes & Taylor (1986) and Taylor & Innes (1988), who have pointed out the increased importance of air breathing at higher temperatures – even *Carcinus maenas* emerges from water to breathe air if the ambient temperature rises above about 28°C. The most complex and efficient crustacean lungs are found in the Trinidad mountain crab, *Pseudothelphusa garmani* (Innes, Taylor & Haj, 1987). Here the lung is not merely a cavity lined with blood vessels, but a series of anastomosing airways that forms a parallel with the vertebrate lung. In water, the branchial chambers are ventilated by the scaphognathites, but in air these movements are replaced by a bellows-like action of the outer parts of the chambers themselves. These movements are intermittent, occurring only every 15 minutes in quiescent crabs, and such intermittent ventilation presumably also reduces water loss. Oxygen uptake is extremely efficient in this system: air flow is uni-directional, there is no 'dead space', and oxygen tension in post-pulmonary blood can reach values very nearly as high as in the airways. This high efficiency is restricted to *Pseudothelphusa*, but many other aspects of respiratory physiology are similar to those in *Holthuisana*

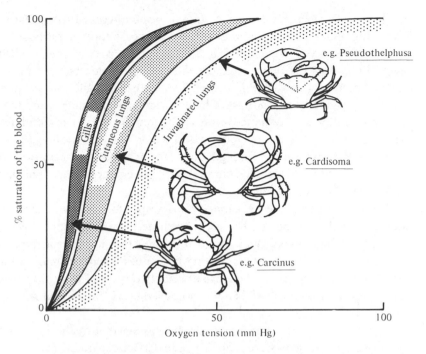

Fig. 9.5. Relation between oxygen affinity of the blood of air-breathing crabs, and type of gas-exchange organ. In aquatic crabs such as *Carcinus*, oxygen uptake through the gills when in air is inefficient, but is aided by high affinity haemocyanin. Values of $P_{50}$ are low. In land crabs with a well vascularized gill chamber, such as *Cardisoma*, gas exchange in air is improved. Values of $P_{50}$ are increased, so that arterial blood is fully saturated after passage through the gill chamber. In the Trinidad mountain crab, *Pseudothelphusa*, the lung can be ventilated by a bellows system, allowing even higher arterial oxygen tensions. In this case $P_{50}$ is higher again. After Innes & Taylor (1986).

*transversa*. As discussed in Sections 4.7.3 and 8.9, both these crabs have close freshwater relatives, and it seems probable that their respiratory systems have evolved quite separately from those found in the marine ocypodids, grapsids and gecarcinids. Innes, Taylor & Haj (1987) speculated that they may have evolved during periods of desiccation near tropical freshwater streams.

Other terrestrial arthropods have quite different varieties of 'lungs'. The two major lines to be considered are the chelicerates (spiders, scorpions and their allies) and the insect – myriapod – onychophora assemblage. In the chelicerates the characteristic respiratory organs are the 'lung books'. These consist of a number of lamellae projecting into an air-filled cavity, which opens to the outside by a pore, the spiracle. Haemolymph circulates through the lamellae, and oxygen uptake occurs across their surfaces. These lung books are thought to be diffusion lungs, although it is possible that some air is ventilated through the spiracle to part of the air-filled cavity. Lung books strongly resemble the external gas exchangers of the aquatic horseshoe crabs (e.g. *Limulus*), which are called gill books. These modified appendages also consist of haemolymph-filled lamellae, and they are protected by an external covering or operculum. Water is circulated over the lamellae when this operculum is raised or lowered. The lung books are thought to have evolved from gill books by invagination, or suturing of the covering plate (Selden & Jeram, 1989), necessitated when the first chelicerates invaded the land.

Many chelicerates possess a second type of respiratory organ, apparently not related to the lung books. These, the tracheae, are found in most spiders together with lung books, but in many arachnids they are the sole type of respiratory organ. They consist of long thin tubes, again opening to the outside through a spiracle. There are at least two types of these tracheae, but, since so little is known of their physiology, they will not be dealt with in detail.

It is in the insects, myriapods and onychophorans that tracheae are best developed. These organs have probably been derived separately from those of the chelicerates, but have a very similar structure. In centipedes, the tracheae are spirally thickened, presumably to ensure that they do not collapse. Air can be flushed through some of these tracheae by body movements, but in the fine branches leading to the tissues – the tracheoles – oxygen moves mainly by diffusion. The spiracles of centipedes and millipedes cannot be closed, so that there is always some water loss by diffusion, but they are usually guarded by complex cuticular lappets that prevent blockage by water and probably also hinder the entry of predators and parasites. Many myriapods living in the soil are exposed to hypoxic conditions, for example when water blocks the air spaces between soil particles. Several species have been shown to accumulate an 'oxygen debt' during this period, and to pay off this debt later (Penteado, 1987).

Tracheal structure and function have been most studied in the insects. The tracheoles penetrate into all tissues in the body, and carry oxygen directly to the cells instead of depending upon an intermediate circulatory fluid. Since the diffusion path in the tissues is not usually more than 10 µm, diffusion produces an adequate oxygen supply (Richards & Davies, 1977), although it is possible that convection currents may also exist (Brocas & Cherruault, 1973). As in myriapods, the tracheae open to the outside through spiracles, but in insects these have specific opening and closing mechanisms. When oxygen is required, as in activity or specifically during flight, the spiracles open, and air is often ventilated into the major tracheal trunks by movements of the body. During rest periods, particularly in dry air, the spiracles close for long periods, reducing water loss, so that respiration becomes, effectively, intermittent.

In most arthropods, gas exchange across the majority of the body surface is minimal because of the impermeable nature of the cuticle. Only in the tracheoles, where the cuticle is particularly thin, can oxygen diffuse rapidly. The general body cuticle, discussed in Section 9.1.2, is water-proofed, and, since water molecules are larger than oxygen molecules, a barrier to water is usually also a barrier to oxygen. In many vertebrates, the situation is more akin to that in the soft-skinned invertebrates as pointed out above, because the skin remains moist and is permeable to gases. Apart from examples such as the eel, however, the skin is usually responsible for only a small part of oxygen absorption in vertebrates. Air breathing in fish and amphibians takes place variously through the lungs, the buccal membranes or through specialized parts of the gut. The selective pressures that may have brought about the evolution of these mechanisms have been discussed in Chapter 8. Although most attention has been paid to the lungfish because of their close phylogenetic connections with the purely terrestrial vertebrates, the teleosts show a much wider variety of structures adapted for air breathing, and are instructive when considering how such mechanisms originated. Many tropical freshwater teleosts spend considerable periods out of water (Dehadrai & Tripathi, 1976). Some have reduced gills, but this loss is compensated for by the vascularization of the buccal cavity, and sometimes by the development of accessory organs in the space above the gill chamber. In *Clarias batrachus*, for instance, arborescent organs have developed on the gill arches (Munshi, 1976), while in *Anabas testudineus* there is a complex labyrinthine organ (Singh & Hughes, 1973). In marine air breathers, such as the gobies of the genus *Periophthalmus* (mudskippers), the gills are again reduced but there is a great increase in vascularized buccal tissue, through which oxygen is obtained (Singh & Munshi, 1968).

It is thought that the early amphibians, emerging from water, may have used buccal respiration to some degree. In present-day amphibians, however, the buccal mucosa is not important in oxygen uptake (Hughes, 1967), and the lungs are the most important uptake site. Oxygen uptake mechanisms in some of the air breathing teleosts give us some clues about how these lungs may have arisen. In particular, there are some species that take up oxygen from specialized areas of the intestine. *Hoplosternum thoracatum*, from South America, swallows air continuously, and absorbs oxygen from the posterior part of the intestine which has an elaborate capillary bed (Huebner & Chee, 1978). Other species such as *Hypostomus plecostomus* do not swallow air in fully oxygenated water, but commence when environmental oxygen tension falls (Graham & Baird, 1982). In the Japanese weatherloach, *Misgurnus anguillicaudatus*, intestinal uptake can account for 20–40% of total uptake, depending upon temperature (McMahon & Burggren, 1987). Some such form of gastro-intestinal breathing could well have been the precursor to development of the lungs as specific evaginations from the gut.

Lungs themselves, which are diverticula originating from a pharyngeal pouch, are widespread in many types of fish as well as in the lungfish and tetrapods. The homology of these various types of lungs has been much disputed, but functionally they are quite similar. Some have developed surfactants – chemicals that line the lung cavities and help to maintain them open. In many teleosts, the primitive lung has been converted into an air bladder and is used as an organ for buoyancy control. The earliest amphibians probably possessed both gills and lungs, as do the present-day lungfish. Whether they could use the lungs when out of water is, however, doubtful. Schmalhausen (1968) has pointed out that, because early amphibian limbs were not braced to the vertebral column by a pectoral girdle, they probably could not support the anterior end of the body off the ground, and inflation of the lungs could not have taken place. Present-day amphibians use a buccal force pump to inflate the lungs, but in higher vertebrates this mechanism has been replaced by aspiration ventilation, using movements of the ribs. It should be noted here that there is still much discussion of the relevance of the present-day amphibians in this sequence. Some authors (e.g. Gans, 1970) believe that recent forms are degenerate and have lost the well-formed ribs of early lines. Others, such as Randall *et al.* (1981) also suggested that aspiration ventilation could have occurred in *aquatic* forms. Clearly there is still much room here for discussion and clarification.

### 9.2.3    Circulatory adaptations and blood pigments

The process of oxygen uptake in the 'lungs' is only the first phase of gas exchange. After this, the oxygen must be transported to the tissues, except in the case of some arthropods, where tracheoles deliver oxygen directly to the cells. In most animals, though, oxygen dissolves in the circulatory fluid and is then unloaded at the tissues. Normally this process is aided by respiratory pigments – haemoglobin in vertebrates and haemocyanin in many molluscs and crustaceans. The question of whether these pigments show specific properties related to aerial, as opposed to aquatic lifestyles has often been discussed. In crabs, for instance, haemocyanin definitely increases the amount of oxygen that can be dissolved in the haemolymph, and its properties ensure that a large proportion of the oxygen is unloaded at the tissues, where oxygen tension is low and carbon dioxide tension is high (Mangum, 1983). One of the variables that define the properties of respiratory pigments is the value $P_{50}$ – the oxygen tension for 50% saturation of the pigment. This value varies greatly from species to species, but cannot be said to relate directly to the degree of 'terrestrialness' of the crabs. Instead, it is related to *internal* ranges of oxygen tension. Indirectly, this is related to the efficiency of the gas exchanger. If this is efficient, internal values of oxygen tension will be high and $P_{50}$ will be high. If the exchange membrane is thick and impermeable, internal oxygen tension will be low, and $P_{50}$ must also be low. The characteristics of the blood pigments must therefore be very closely integrated with the total respiratory mechanism. In some groups, such as the example given for crabs in Fig. 9.5, there appears to be a clear cut trend from water to land, but efficiency of oxygen exchangers may also have much to do with the general activity of the species, and such series are far from universal.

The circulatory system, in contrast, has changed in much more uniform ways from aquatic to terrestrial animals. These changes are perhaps best understood in the vertebrates. In water breathing fish, blood is pumped by the heart directly through the gills. Aerated blood then passes through the dorsal aorta to the body tissues. In the lungfish, with gills *and* lungs, the blood pumped by the ventricle can essentially be separated into two streams (Burggren & Johansen, 1987). One passes through the anterior gills in the normal fashion, but blood passing through the posterior gills can pass to the pulmonary artery. This second route is accentuated when external oxygen levels fall (Johansen *et al.* 1968). Since oxygenated blood from the lung returns immediately to the heart, it might be thought that this mechanism would be very inefficient, because the oxygenated blood would mix with de-oxygenated blood from the body. However, a spiral valve in the conus arteriosus effectively separates the two bloodstreams in the heart, and oxygenated blood is passed through the gill arches to the body while de-oxygenated blood passes to the lung. This separation of pulmonary and systemic circuits also occurs in amphibians, although there is little anatomical change there (Johansen & Hanson, 1968). With the transfer of all respiratory exchange from skin and gills to the lungs, as in the reptiles, further separation of pulmonary and systemic circuits evolved. Complete anatomical separation is found in the birds and mammals – the so-called double circulation.

### 9.2.4    Carbon dioxide removal

In aquatic animals, the elimination of carbon dioxide routinely occurs across the body surface or through the gills. Since little is known of the mechanisms by which carbon dioxide is eliminated on land in many groups, we will consider only the crabs and the vertebrates, which provide surprisingly similar pictures. Aquatic fish and aquatic crabs excrete their respiratory carbon dioxide through the gills. Carbon dioxide is very soluble in water, and it readily passes across the gill membranes, so that the carbon dioxide tension in the blood remains low. In amphibious fishes, however, the gills are often exposed to air, and cannot therefore excrete carbon dioxide. In many amphibious and air breathing fish this responsibility has been moved to the skin, where carbon dioxide diffuses into the air. In the electric eel, *Electrophorus electricus*, 81% of the carbon dioxide is lost across the skin (Farber & Rahn, 1970). In present-day amphibians, this 'bimodal' form of breathing is also present: the lungs take up 50–78% of the oxygen needed, but 57–84% of the carbon dioxide produced is lost through the skin (Rahn & Howell, 1976). Some lungfish can excrete part of their carbon dioxide across the skin, while others, e.g. *Neoceratodus*, cannot do so. The gills are the normal route for carbon dioxide elimination, even in *Protopterus*, which in water loses more than twice as much through the gills as through the lungs. This inability to lose carbon dioxide through either gills or lungs when in air is one of the major constraints preventing emergence from the water.

Air breathing crabs represent a parallel with the lungfish. They cannot eliminate carbon dioxide through the hard carapace, but must maintain their gills moist and use them to pass carbon dioxide into water (Innes & Taylor, 1986). By this means they can maintain low carbon dioxide tensions in arterial blood, but at the expense of retaining a connection with water. In the robber crab *Birgus latro*, most of the carbon dioxide is eliminated via the gills when resting, but in activity about half is lost via the lungs (Greenaway, Morris & McMahon, 1988).

Truly terrestrial vertebrates, on the other hand, have impermeable skin, and no gills, and *must* lose their carbon dioxide through the lungs. This became possible with improvements in the mechanics of pulmonary breathing, allowing air to be flushed out of the lungs more efficiently. Together with this change came another – the site of the carbon dioxide receptors. In fish these are external, since the important variable to be measured is environmental carbon dioxide: internal carbon dioxide tensions are always low. In terrestrial vertebrates, the carbon dioxide receptors are internal, and monitor tensions in the blood. This level can then be changed by changes in the rate of ventilation of the lungs. A discussion of this subject is given by Dejours (1988).

One parallel among the crabs can be compared with the terrestrial vertebrates. This is the Trinidad mountain crab *Pseudothelphusa garmani*, already mentioned in Section 9.2.2 (Innes & Taylor, 1986). Because the lungs in this species are ventilated by a bellows action which provides a rapid one-way flow, gas exchange can be exceedingly efficient. The diffusion distance is only 0.4 μm, and the rapid flow of air through the system effectively 'blows off' the carbon dioxide: carbon dioxide tensions in crab haemolymph remain low, despite the fact that the animal can breathe only through the lungs.

### 9.2.5    Acid-base balance

Variable levels of carbon dioxide in the blood lead inevitably to the problems of acid-base balance. In fish, as described above, carbon dioxide tensions in the blood are always low. The regulation of blood pH is carried out at the gills, where $Na^+ - H^+$ and $Cl^- - OH^-$ pumps operate (Cameron, 1978). Although fish kidneys are, to some extent, involved in the excretion of hydrogen ions (Wood & Caldwell, 1978), they are not involved in regulating respiratory acidosis. Even for other substances that might affect blood pH, such as ammonia, the kidneys of fish are much less important than the gills. Overall, then, acid-base balance in fish is controlled by the gills.

In lungfish, excretion of carbon dioxide and ammonia still occurs through the gills in water. On land, however, this cannot occur, and carbon dioxide levels build up internally. Carbon dioxide tensions in lungfish out of water are therefore higher than in aquatic fish. It is this higher carbon dioxide tension in the blood that promotes some diffusion through the skin. The ammonia, however, is mainly converted to urea, and stored within the body.

This situation evidently provides no long-term solution to life out of water. Although lungfish can aestivate on land for long periods, they must in the end return to water. The amphibians face similar problems, although they have some further means of acid-base control. They have a very well vascularized skin, and lose much carbon dioxide directly to the air. In the skin they also have exchange pumps for $HCO_3^- - Cl^-$, so that they can regulate pH when in water. Another difference from the lungfish is found in the kidney, which now takes some part in pH regulation: as carbon dioxide tension rises in the blood, the concentration of $HCO_3^-$ is increased, probably by renal excretion of $H^+$ ions. Perhaps the most significant difference between amphibians and fish is, however, the ability of some toads to regulate carbon dioxide tension in the blood by changing the rate at which they ventilate their lungs (Boutilier *et al.*, 1979). It is *this* method which has been greatly developed in higher tetrapods, and in mammals it is the major method of respiratory acid-base control.

**9.3    Nitrogenous excretion**

The excretory organs responsible for eliminating nitrogenous waste products from the body have already been discussed with reference to water balance in Section 9.1. Until relatively recently it has been considered that in general these organs excrete the toxic product ammonia from aquatic animals – where it can be rapidly dispersed in water – but urea or uric acid from terrestrial animals because these are relatively non-toxic or insoluble respectively. Such a generalization has now been shown to be much too over-simplified (Little, 1983). This section will summarize what is known about four animal groups – gastropod molluscs, isopod crustaceans, insects and vertebrates – to show the enormous range in excretory mechanisms, and to show the complex relationship of excretory products to the environment.

Prosobranch gastropods excrete a variety of nitrogenous compounds (Little, 1981). Many marine species show high concentrations of uric acid, especially in tissues surrounding the blood sinuses, but also possess the enzymes capable of breaking down uric acid to ammonia. Amphibious freshwater snails such as *Pomacea*, and their terrestrial relatives the cyclophorids, also accumulate uric acid and other purines in part of the kidney, as well as producing some urea. The relation between nitrogenous end product and habitat is far from clear, but the ability to *store* uric acid may be greater in amphibious and terrestrial forms. Perhaps this temporary measure is more necessary in the changing terrestrial climate than in the sea. A further ability, found in aestivating terrestrial prosobranchs, is shared by the terrestrial pulmonates: these can excrete gaseous ammonia through the mantle epithelium (Speeg & Campbell, 1968). To excrete ammonia on land might seem a surprising faculty, but by doing so the snails prevent the build-up of excretory wastes, and avoid the problems of storage. Although ammonia is toxic, the process of excretion in aestivating pulmonates allows it to form only at the mantle surface, in conjunction with the deposition of calcium carbonate for shell formation. This process allows *Helix aspersa* to eliminate as much as 30% of its waste nitrogen directly into the air.

The isopod crustaceans show some parallels with the pulmonate gastropods. All isopods are, however, primarily ammonotelic, and in aquatic forms ammonia rapidly diffuses out of the body. Terrestrial woodlice also release the major part of the waste nitrogen as ammonia, but as a gas (Wieser, Schweizer & Hartenstein, 1969). This ammonia is probably evolved from the surface of the pleopods, but the exact mechanism has not been investigated. Ammonia release does not occur continuously (Fig. 9.6). Maximum rates occur in the daytime, and it has been proposed that this allows the woodlice to lose nitrogen without losing water, because at this time they remain in humid micro-habitats. When foraging for food at night, where they run the risk of desiccation, little ammonia is released. Since a similar pattern is found even in marine littoral isopods, it has probably evolved in the littoral zone and has acted as a pre-adaptation for terrestrial life.

The insects conform to some extent to the classical theory, since many aquatic larvae excrete ammonia, while many terrestrial adults excrete uric acid (Stobbart & Shaw, 1974). The picture is more complicated than this, however, because the majority of terrestrial pterygote insects excrete a mixture of uric acid, allantoin and allantoic acid. Each of these have different solubilities, and the proportions excreted may be related to different lifestyles. In *Dysdercus fasciatus*, for example, allantoin is the major end product. This species has a high rate of urine production, and allantoin, which is ten times more soluble than uric acid, is stored in the haemolymph before its excretion. Further complications have been detected by recent studies, even in the cockroach (Mullins & Cochran, 1973). Cockroaches were thought to be uricotelic, but in fact the American cockroach excretes primarily ammonia and amino-nitrogen compounds. Evidently there is a need for a much more detailed and wide-ranging study of excretion in the insects before meaningful generalizations can be made.

The situation in the vertebrates is as complex as in the invertebrate groups. The ability to form urea appears to have evolved in the early fish (Smith, 1959), although we are not certain of the selection pressures involved. Almost all present-day fish groups can produce both ammonia and urea. The lungfish *Protopterus* excretes ammonia and urea in approximately equal proportions when in water. Both are excreted through the gills, so that in air nitrogen cannot be eliminated. Instead, all waste nitrogen is converted to urea which is then stored in the tissues. A similar pattern is found in the mudskippers, *Periophthalmus*: these are ammonote-

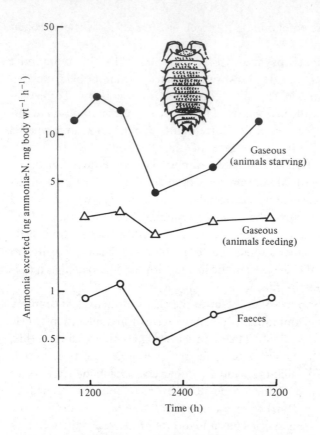

Fig. 9.6. Excretion of ammonia by the woodlouse *Porcellio scaber* in relation to time of day. The majority of ammonia is released as a gas, and only a small fraction is contained in the faeces. Gaseous release occurs mainly during the daytime, when the woodlice are in humid micro-climates, so that evaporation of ammonia would not be accompanied by evaporation of water. After Wieser & Schweizer (1970).

lic in water, but switch to urea production in air (Gordon, Ng & Yip, 1978). Once again, excretion occurs through the gills so that urea must be stored when out of water. In the amphibians, in contrast, which also excrete a mixture of ammonia and urea, waste nitrogen is excreted by the kidneys. This is a step of immense importance, because amphibians have no limit set by nitrogen excretion to the time they can stay in air.

The reptiles and birds show an immediate contrast with the amphibians. Their major excretory product is uric acid, and this has for long been regarded as an adaptation to reduce water loss during excretion. While this may be true in reptiles, birds excrete uric acid as a colloidal gel containing large quantities of water, and Sykes (1971) has calculated that they lose as much water by excreting this gel as the mammals lose by excreting urea. It seems more likely that in birds the excretion of uric acid evolved to cope with the accumulation of nitrogenous waste in the cleidoic egg, and that in part, at least, the excretion of uric acid by adult birds is a retention of this embryonic adaptation. In reptiles, however, the embryos produce mainly urea, not uric acid, while it is the adults that excrete mainly uric acid, suggesting that here the excretion of an insoluble end product is an adult adaptation. Coupled with the ability of some amphibians to produce uric acid, these differences suggest that the vertebrate excretory pathways are extremely labile, and that the production of uric acid has evolved independently several times, probably under different selective pressures.

## 9.4 Reproduction

In many marine invertebrates, sperm and eggs are liberated into the sea, and fertilization is very much a random process, although the likelihood of fertilization may be improved by synchronized spawning. The fertilized eggs often develop into pelagic larvae which utilize a food source quite different from that of the adults. Such a reproductive pattern is inappropriate for life on land, and is also rare in fresh water. It may also be inappropriate for many kinds of lifestyle in the marine littoral zone, since many animal groups on the shore show internal fertilization and produce large eggs which hatch as juveniles rather than as pelagic larvae. Some also show parental care in the form of viviparity or brooding. These features may all be considered to be possible pre-adaptations to life on land, and in some cases the reproductive features of terrestrial lines have been directly inherited from intertidal ancestors. The same can be said for some freshwater adaptations, which are present in terrestrial animals of freshwater ancestry. It is important to note, however, that such apparent pre-adaptations are adaptive for particular species in their own niches, and *some* are *not* found in terrestrial relatives. For example, several marine nemertines show internal fertilization and viviparity, but terrestrial species show no care of the young.

The fact that reproductive patterns are geared to the details of microhabitat and lifestyle is well exemplified by the variety of reproductive strategies found in marine intertidal littorinid gastropods. In Britain there are eight species of Littorinidae, found from low water of spring tides up to regions well above the high water mark. These show a variety of methods by which juvenile snails are produced, as discussed in Section 7.4. Four species lay benthic egg masses, two are ovoviviparous and two release planktonic egg capsules. Yet these methods do not appear to be related to height on the shore, or to degree of independence from the sea. The two ovoviviparous species *L. neglecta* and *L. saxatilis* are widely distributed and *L. saxatilis* may be found very high on the shore; but so many *L. arcana*, which lays benthic egg masses. *Melarhaphe neritoides*, which produces planktonic egg capsules, is found *highest* on the shore. There is, therefore, no trend in reproductive method from low water to the supratidal that might allow us to infer which particular aspects of reproduction are of critical importance in the invasion of land. In fact, terrestrial littorinaceans such as *Pomatias elegans* show internal fertilization like the marine littorinids, but are not ovoviviparous. They lay egg capsules in leaf litter and soil, and the eggs are relatively permeable so that they require a specific composition of external water for development.

Many of the terrestrial crustaceans, unlike the gastropod molluscs, show care of the young. This capacity for broodcare is universal in marine amphipods and isopods, and is retained in terrestrial lines, so that the young are released as miniature adults, having up to this stage been protected from desiccation, disease and predation, and supplied with food by the mother. Other reproductive aspects have changed considerably from aquatic to terrestrial lines. In aquatic amphipods, the male carries the female below him, dorsal surface up, for a long period before copulation occurs. Egg laying takes place immediately after copulation. In semi-terrestrial species the carrying period is very short. The males carry the females lateral surface up and tranversely, so that they can still walk and jump effectively (Williamson, 1951). Egg laying may occur some considerable time after copulation, since it is governed by the time of the moult.

Care of the young is also found in most terrestrial crabs (Powers & Bliss, 1983). Copulation often occurs in burrows, and in freshwater families the females do not emerge from these burrows until the young crabs have hatched and are held beneath the abdomen by the pleopods. The young are then released into water by the mother and spend some time there before returning to land. The grapsids and ocypodids, which are mostly marine in origin, release their larvae into the sea, and even *Gecarcinus lateralis*, possibly the most terrestrial crab, has to return to the sea to release its zoea larvae (Bliss, 1979). The involvement of pelagic larvae in the life cycle of these crabs of marine origin limits them irrevocably to a maritime existence. Although *Gecarcinus* can be found many kilometres inland, it must make its annual migration back to the sea. The freshwater crabs, in comparison, can be found at high altitudes and at great distances from the sea, but always near a source of fresh water. In both cases, then, the possession of a

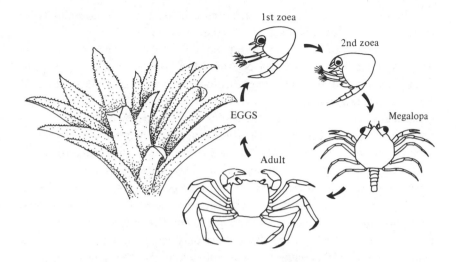

Fig. 9.7. The life cycle of a grapsid crab, *Metopaulias depressus*. This crab lives and breeds in large epiphytic bromeliads in Jamaica such as *Aechmea paniculigera*, which retain reservoirs of fresh water between the leaves throughout the year. The female crabs carry the eggs until they hatch as first zoea larvae which rapidly moult to second zoeas. Unlike the larvae of marine crabs, neither of these larvae can swim freely, and they do not feed. The megalopa is a crawling stage, and feeds actively, subsequently metamorphosing into a juvenile. After Hartnoll (1963).

pelagic larval stage has made them much more limited in distribution than the amphipods and isopods, which need neither salt nor fresh water to breed.

There have, however, been some significant changes in the mating procedures of crabs in their movement out of the sea. Most crabs can copulate only when the female exoskeleton is soft, just after she has moulted. In the grapsid *Sesarma*, for instance, the genital operculum is fused in a closed position except after the moult. More 'terrestrial' crabs are able to copulate when the female is hard: in the grapsid *Metopaulias depressus*, which lives in epiphytic bromeliads in Jamaica (Hartnoll, 1963), the female genital operculum is hinged so that both copulation and egg laying can occur at any phase in the moult cycle. The life cycle of this species is shown in Fig. 9.7. A second significant change concerns courtship behaviour, and is well exemplified by the ocypodid crabs. Courtship is of very variable complexity in the genera *Uca* and *Ocypode*, the fiddler and ghost crabs, which provide sequences from simple claw-waving followed by direct copulation, to complicated displays and the building of specialized copulation burrows. The general increase in behavioural complexity in more terrestrial crabs is discussed further in Section 9.7.

A completely different set of reproductive adaptations is shown by arthropods of the insect-myriapod line. Many of the primitive groups in this assemblage belong to the cryptozoic fauna, and reproduce in soil and leaf litter, in an essentially interstitial environment. In both chilopods (centipedes) and symphylans, the males deposit sperm droplets or spermatophores, either directly on the soil or on specially constructed webs or threads. Female centipedes take up the sperm through the genital opening, while female symphylans simply eat it. The presence of copulation in diplopods (millipedes) appears to be a secondary adaptation. Sperm droplets and simple stalked spermatophores are also the rule in the apterygote insects – the Thysanura, Diplura and Collembola. In some cases these are apparently found by the females only accidentally, but in others a complex courtship behaviour leads the female to the spermatophore. According to Schaller (1979), the presence of spermatophores or sperm droplets is probably a primitive feature that arose in the interstitial habitat before the insects or myriapods invaded land (see 5.2.7). Here the discharge of gametes into open water was not possible, and the use of stalked spermatophores provided a pre-adaptation allowing sperm transfer in the humid cryptozoic habitats on land. In the pterygote insects that have emerged from the cryptozoic habitat, the use of sperm droplets has been replaced by copulation and internal fertilization.

The vertebrates demonstrate yet another approach to reproductive adaptation, influenced by their evolution in fresh water. The eggs of freshwater fish are larger than those of marine species, usually in the range of several millimetres diameter as compared with less than 2 mm (Wootton, 1979). This larger egg size is probably necessary in order to provide the embryos with a large food supply in an environment where zooplankton can fluctuate enormously in abundance and at times is very scarce. It also reduces the surface area: volume ratio, and therefore the osmotic work that needs to be done. A larger egg size therefore relates to a larger size of the larva at the time of hatching, with all its advantages. If the adoption of a larger egg size in terrestrial vertebrates is indeed related to their freshwater ancestry, it is interesting to speculate that, had the terrestrial vertebrates evolved in marine environments, their reproductive adaptations might have been quite different. The retention of small pelagic eggs, for instance, would have left the group with an irrevocable tie to the sea, as found in many of the present-day crabs.

The amphibians have retained a large egg size, but the maximum diameter of present-day amphibian eggs is restricted to about 9 mm (Carroll, 1970). Above this size diffusion would not allow a sufficient oxygen supply to reach the tissues. The reptiles, birds and mammals have evolved accessory breathing structures in the egg that have allowed egg size to increase further. The chorioallantoic membrane, a well vascularized structure, lies below the outer egg membranes, so that oxygen diffusing in through the egg shell can be absorbed and transported to the embryo by the blood system. Just when this structure evolved, in relation to the protective shell, is not known, nor are the selective pressures under which it developed. Discussion of these points has been given by Szarski (1976) and Luckett (1976).

**9.5    Support and locomotion**

In soft-bodied animals, body turgor is maintained by the action of the muscles in the body wall, usually both circular and longitudinal. The pressure that they exert on the internal fluids, often a coelom, creates what is known as a hydrostatic skeleton: since the fluid is virtually incompressible, changes in tension of specific sets of muscles create changes in body shape. In water, this system is common, because the high specific gravity of water, together with its viscosity, give external support to the body. In air, this support is lacking, and animals like jellyfish and octopuses could not maintain their normal body shape. All terrestrial soft-bodied animals therefore have thick layers of body muscles to maintain themselves erect, but even then they are generally restricted to small sizes: large animals with a hydrostatic skeleton cannot obtain enough support in air.

Among the smallest of the animals considered in this book are the nematodes. They live essentially in the water films between soil particles, and move by contracting their longitudinal muscles to form a swimming motion, possible only in water. Some of the larger soft-bodied animals show some peculiar adaptations to moving on land. Flatworms, for instance, move by muscular waves on the ventral surface, and those on land form waves so that the body touches the ground only at intervals, leaving 'footprints' of mucus behind, and not losing as much water as would happen with continuous contact with the ground (Fig. 9.8). The nemertines normally move by ciliary gliding, again in mucus, but as an escape reaction some of the semi-terrestrial and terrestrial species can evert their proboscis, and, by adhering to the ground with the tip of this, the body can be drawn rapidly forward by contraction of the longitudinal muscles.

The commonest animals on land with hydrostatic skeletons are the annelids and molluscs. Earthworms move by alternately contracting and

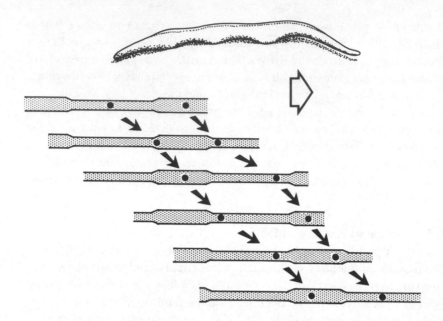

Fig. 9.8. Locomotion in a terrestrial flatworm. The top diagram shows a species from New Guinea, in which the body touches the ground in only three places. The lower diagrams show how locomotion may be produced by a combination of ciliary gliding and muscular peristaltic waves. The peristaltic waves move backwards along the animal, while the animal glides forwards using cilia at the places where the body contacts the substrate. The result is that the waves remain stationary relative to the substrate, while individual points on the animal move forwards, but not at a regular speed: two such points are shown by the black dots, movement of each being shown by the arrows. The reduction of contact with the ground probably reduces water loss. After Jones (1978).

expanding the body segments, and this mechanism works well in the soil where the physical structure is little different from that in aquatic sediments. More active in the open are the leeches. These have suckers at each end of the body, and have lost all internal septa. The body cavity is filled with mesenchyme cells which form a hydrostatic skeleton, and leeches can move very actively by stretching the body forward, anchoring with the anterior sucker, and then drawing up the posterior one behind. Their great mobility allows them to seek out and attach to vertebrate prey in tropical forests.

The gastropod molluscs move on a flattened foot by muscular waves, in much the same way as aquatic species. A unique specialization for rapid movement is found in the Pomatiasidae, where the foot is divided longitudinally. Each side of the foot performs alternate stepping movements (Lissmann, 1945). Some of the other gastropods can move so that only parts of the foot touch the ground, and they leave 'footprints' reminiscent of the flatworms.

Finally, among the soft-bodied animals, it is appropriate to mention the Onychophora. Although these are often regarded as arthropods, their cuticle is flexible like insect caterpillars, and they utilize a hydrostatic skeleton. The body muscles maintain rigidity, but the motive force is provided by the extrinsic muscles of the legs. Here, then, they resemble animals with a hard exoskeleton. *Peripatus* spp. can change their shape immensely, and are specialized for life in the cracks and crevices of rotting logs.

Most arthropods on land move using the same principles as their aquatic ancestors: the limbs articulate at specific joints, and each has a set of flexor and extensor muscles. The chelicerates are peculiar here, possessing only flexor muscles in the legs, and extension is brought about by an increase in hydrostatic pressure. Speeds in terrestrial arthropods are generally higher than in marine species, and there are some interesting modifications in movement mechanisms found in terrestrial crabs. Of these crabs the grapsids reach 1 m.s$^{-1}$, but the fastest species are found in the ocypodids. *Ocypode ceratophthalma* can run at well over 2 m.s$^{-1}$, can accelerate rapidly, and can reverse directions without losing speed. When moving at moderate speed, two pairs of legs alternate on each side, so that the crab is effectively 8-legged – the chelipeds are not involved. At maximum speed, only 3 legs touch the ground: two legs provide the thrust and one acts as a skid to provide balance (Burrows & Hoyle, 1973). In this mode the crab in fact leaps through the air so that its step length is greater than it could be when walking.

The remainder of the arthropods provide a great variety of locomotory gaits and mechanisms, and these cannot be considered in detail here, particularly as we have no evidence about the mechanisms used by their aquatic ancestors. The myriapods in particular show a great diversity (Manton, 1979). Most millipedes are specialized for burrowing, and move slowly but with great force. Centipedes, on the other hand, are generally specialized for running, and of these *Scutigera coleoptrata* is perhaps the fastest. It has greatly elongated legs which have a short back-stroke and long forward (recovery) stroke (Fig. 9.9). Although it has negligible pushing power, it can move at 42 cm.s$^{-1}$.

The insects show more variety of mechanisms of movements, from walking to running, jumping and flying. Some of the apterygotes such as

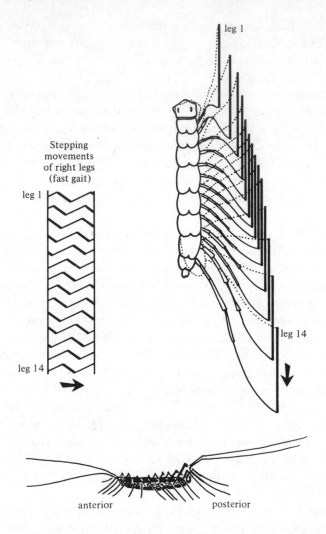

Fig. 9.9. Locomotion in the centipede *Scutigera coleoptrata*. The lower diagram shows the animal standing. Its long antennae (left) are sensory, as are the last pair of legs (right). The upper right diagram shows the movements of the 14 legs. Solid lines show positions at the end of the propulsive stroke; dotted lines show positions at the end of the recovery stroke. Note that the body moves laterally to some extent (solid and dotted lines). The diagram on the left shows the stepping movements of the right legs and the phase difference between legs in the fastest gait. Thin lines show recovery stroke, thick lines show propulsive (back) stroke. After Manton (1952a,b).

the Thysanura and Collembola move essentially by jumping, the latter by use of a specialized forked springing organ which can be held beneath the body and then suddenly straightened; hence the name 'springtails'. Undoubtedly the most critical development in insect locomotion has been the evolution of flight, but this is outside the scope of this book.

The vertebrates provide the most complete sequence of intermediates between aquatic and terrestrial locomotion (see also Section 8.3). Typical fish swimming is based on contraction of the myotomal musculature and oscillation of the tail fin, to provide thrust, while steering, braking and manoeuvring are effected by the paired fins. In many fish, including the teleosts, buoyancy is aided by the swim-bladder. A major departure from this scheme is the use by some fish of the fins as walking limbs or paddles. Such adaptations are seen in a variety of present-day teleosts, like the gurnards, but the first known development in this direction occurred in the rhipidistians, the reputed ancestors of the amphibians. In these Devonian fish, the pectoral and pelvic fins were supplied with their own axial skeletons, allowing their use for support and for movement over surfaces, and in consequence they have been called 'lobe-fins'. These fins were able to support the fish only *under* water, because they had no direct articulation with the main axial skeleton. A similar organization is found in present-day lungfish, and in the marine coelacanth, *Latimeria*.

There is a gap in the fossil record between the rhipidistians and the earliest amphibians, which makes it difficult to follow the exact development of limbs out of water. The earliest amphibians known in detail were the Ichthyostegalia from east Greenland. In *Ichthyostega*, each limb terminated in individual digits. The limbs connected with relatively large pectoral and pelvic girdles, and these in turn were attached to the vertebral column. In spite of these connections, though, it seems unlikely that *Ichthyostega* could have supported the body off the ground when out of water. By comparison with present-day amphibians, it has been concluded that *Ichthyostega* moved on land at least partly by using the myotomal muscles: these produced a sinuous bending movement as in fish, with the limbs doing little except keeping the body upright and providing anchoring points. Edwards (1976), from an examination of the locomotion of present-day salamanders, has concluded that the sinuous bending of the body, in which the pectoral and pelvic girdles rotate in the lateral plane, first one way and then the other, was the primitive method of movement. This method is shown in extant salamanders when they are in the 'trotting' gait. When in the slower 'walking' gait their body is held more rigidly, and most propulsive force is generated by active retraction and protraction of the limbs. This method seems to have arisen as a later development, made possible by the ability to rotate the limbs, rather than the girdles. It should be noted, however, that some early labyrinthodonts probably did have the ability to rotate the limbs actively, and it has been proposed by Rackoff (1980) that active limb protraction and retraction evolved *before* movement on to land.

The evolution of locomotory mechanisms in reptiles, birds and mammals shows enormous increases in the ability to support the body off the ground, brought about by lessened dependence on body musculature. Great increases in both speed and manoeuvrability have developed, and in parallel with the insects the evolution of flight has vastly enlarged the niches potentially available to birds.

### 9.6    Food supply

In the sea, the most widely available food source is probably the plankton, and this is reflected in the abundance of filter feeders. At least five marine phyla consist *entirely* of filter feeders – Porifera, Entoprocta, Phoronidea, Bryozoa and Brachiopoda. In fresh waters, plankton may also be abundant, if more variable in supply, and the Porifera and Bryozoa are common, together with the Rotifera which are scarce in the sea. Brachiopoda are not found in fresh water, and Entoprocta are rare. Although there *is* aerial plankton above the surface of the land, and drifting dead organic matter, the concentrations are low, and appear to have been unable to support a filter-feeding lifestyle on land.

Animals living on land are therefore either detritivores, herbivores or predators, or some mixture of these. Surprisingly, about half the major terrestrial groups are predators, while the other half comprise the lower levels of food chains. Flatworms, nemertines, leeches, centipedes, onychophorans, amphibians and reptiles are *primarily* predators. The earthworms, gastropods, isopods and amphipod crustaceans, millipedes and insects, possibly with the birds and mammals, can be classified primarily as herbivores or detritivores. The ways in which these various groups interact are discussed in Chapter 10, but here a comment may be made on two points. First, one animal-plant relationship lacking on land is the dependence upon symbiotic algae found commonly in the sea and fresh water: there are no terrestrial equivalents of the zooxanthellae found, for example, in corals and bivalves. Secondly, it is possible that the relatively small number of herbivorous animal groups on land may relate to the nature of the terrestrial food supply. In the sea, most herbivores depend upon microalgae, which have low carbon:nitrogen ratios and fast growth rates. On land, the dominant available food source consists of macrophytes, which have high carbon:nitrogen ratios and low growth rates, and which contain much structural material of low nutritive value. As pointed out in Chapter 10, the evolution of terrestrial herbivores has therefore necessitated different approaches from those of aquatic lines.

There is little evidence to suggest how, if at all, animal lines changed their food sources as they moved on to land. Essentially it seems that aquatic predator lines remained predatory on land. This is so for flatworms, nemertines and leeches. In the vertebrates, the aquatic rhipidistians were predators, and present-day amphibians and reptiles have mostly remained so. Many past reptile lines have become herbivorous, and present-day birds and mammals have diversified into almost all trophic niches. In some groups, such as the prosobranch gastropods, it is the herbivore groups that have invaded land, while predators have never left the sea. This may be because most predatory marine gastropods, the Neogastropoda, feed on sessile invertebrates, and require long periods to attack their prey. Such immobile animal food items are scarce on land.

Herbivores and detritivores have been very successful in terrestrial ecosystems. The earthworms, gastropods, isopods and amphipods are major consumers of detritus and living plants, and the insects in particular have diversified to take advantage of the variety of flowering plants. These topics are discussed further in Chapter 10.

## 9.7    Sense organs and behaviour

Several aspects of behaviour have already been considered – those concerned with water loss (9.1.6), reproduction (9.4) and locomotion (9.5). There still remain, however, two senses intimately related to the properties of air and water, and the relationships between these and behavioural patterns. These two senses are hearing and vision.

### 9.7.1    Hearing and sound production

Many aquatic invertebrates possess some vibration sense, and some of the aquatic crabs can stridulate. It is not known whether such aquatic low-frequency vibrations can be detected only through the substrate, or whether they can be 'heard' after travelling through water. In general, however, the sense of hearing – the ability to detect high frequencies – appears to be restricted to those animals active in air. In the marine intertidal, hearing is best documented in the ocypodid crabs (Salmon & Hyatt, 1983). *Uca* species can make noises by rapping on the ground with their chelipeds, but these noises are transmitted through the substrate, and detected as low-frequency vibrations. *Ocypode* species, the ghost crabs, can make noises by stridulating – rubbing the tubercles on one part of the cheliped against ridges on another part. These air-borne sounds can be detected by myochordotonal organs on the legs (Horch, 1971). Stridulations are used by males to attract females to the burrow, and this communication between the sexes seems to be the major use of sound production in the invertebrates. The males of some terrestrial insects, such as the crickets, grasshoppers and cicadas, have a well-developed capacity for sound production, and their sounds are used to attract females: both males and females have specialized sound receptors. The detection of sounds is even more widely used by the vertebrates: sounds give warning of predators, are used to detect prey, and are used in communication and courtship.

### 9.7.2    *Vision*

The ability to detect light is virtually universal in the animal kingdom, but the sensitivity of so-called 'eyes' varies from those that can differentiate between various light intensities to those that can form an accurate image of the external world (Autrum, 1979). Because visual communication is more effective over long distances in air than in water, it might be concluded that visual systems and visually oriented behaviour would be more complex on land. With the knowledge that the cephalopod and vertebrate eyes have remarkably similar structures and abilities, however, this cannot be taken as a generalization. Taking the crabs as an example, it must be admitted that vision *seems* to play a greater part in the lives of those such as *Ocypode*, which are active in air, than in subtidal forms. Similarly, in the amphipods, the sand-hopper *Talitrus* is reputed to be able to distinguish objects some metres away, whereas aquatic amphipods certainly cannot do this. Some of the intertidal amphipods are also thought to be able to navigate using the plane of polarization of light (Herrnkind, 1983), and the same may be true of intertidal gastropods, whereas aquatic forms seem to have no such ability. Nevertheless, it must be stressed that in part this difference is only an apparent one. Until we know more of the sense organs of aquatic animals, the idea that vision on land is better than vision in the water must be regarded as a gross over-simplification.

### 9.7.3    *Behaviour*

The complexity of behaviour patterns is also often thought to have increased from aquatic to terrestrial animal lines. For example, the contrast between the aquatic crabs such as *Carcinus maenas*, which hunt for their prey using chemical and tactile cues, and *Pachygrapsus crassipes*, which hunts using visual cues, is dramatic. Not only is *Pachygrapsus* more active, but it seems more 'aware' of its surroundings. Similarly, the complexity of social interactions in the fiddler crabs, *Uca*, seems to increase from low-shore to high-shore species: the low-shore species have rhythms strictly geared to the tides, while the high-shore species are active for much longer periods, and show complex courtship displays. One reason for this may be that movement can be faster in air, and with increased speeds sense organs *need* to be more rapid-acting. This may in turn lead to the possibility of more complex interactions. If this sequence is a genuine one, it would fit well with the points made in Chapter 2 about terrestrial vs aquatic habitats. Since terrestrial habitats show more spatial and temporal variation than aquatic ones, it would intuitively be *expected* that behaviour would be more complex on land. However, as with the subject of vision, it is important to be cautious here. Our knowledge of the normal behaviour of aquatic animals is only now beginning to show how complex social interactions can be under water. Until a comparable amount of information is available about aquatic and terrestrial representatives of the same animal lines, even the generalization that terrestrial behaviour patterns are more complicated than aquatic ones may be an over-simplification.

## 9.8    Conclusions

Having detailed some of the adaptations to life on land found in present-day animals, it is evident that there have been several successful approaches to an invasion of terrestrial conditions. Many of the smaller animal types have invaded cryptozoic habitats, and in so doing have adopted relatively few physiological modifications. Terrestrial flatworms, nemertines and amphipods, and the soil-dwelling nematodes and protozoans, show no marked physiological or structural 'advances' when compared with aquatic relatives. Theirs has been an essentially behavioural adaptation to life on land.

An alternative approach has been to maintain activity in moist conditions, but to withstand total desiccation by the process of cryptobiosis. This procedure has been adopted by several animal lines of very small size: presumably the rotifers and tardigrades, with their large surface area:volume ratio could never physiologically retain water in the face of a desiccating atmosphere, yet the evolution of complete tolerance of water loss has allowed them to invade even temporarily damp areas.

Most other terrestrial animals show some degree of physiological adaptation to life on land. Even the essentially cryptozoic fauna such as the woodlice, or the hygrophilic gastropods, show a combination of behavioural reactions, which maintain them in favourable micro-habitats, with physiological ability to withstand some desiccation, and to modify respiratory and excretory processes depending upon external conditions. Many of the arthropods and vertebrates have carried these physiological abilities so far that they are able to exist in desert climates. In contrast with adaptations for water balance, other terrestrial modifications seem to show surprisingly little relation to the routes adopted to invade land. Respiration and excretion, for instance, show similar types of adaptation in animals with marine and freshwater ancestors, and this is presumably because the selection pressures for these processes were similar in the two environments.

Several aspects of animal life-processes seem to have changed little in the movement on to land. Locomotion has altered drastically in the vertebrates, but not apparently in other groups. Feeding mechanisms show a similar variety on land to those found in water, with the exceptions pointed out in Section 9.6. Behaviour may be more complex on land, but even this is not well established. Nevertheless, terrestrial ecosystems are, in many ways, different from aquatic ones. The reasons for this are now discussed in the final chapter.

# 10

## Terrestrial ecosystems

It is difficult to pronounce which division of the earth, between the polar circles, produces the greatest variety. The tropical division certainly affords those which principally contribute to the more luxurious scenes of splendour . . . But the temperate zone exhibits scenes of infinitely greater variety.

William Bartram (1791) *Travels through North and South Carolina, Georgia, East and West Florida*, Philadelphia: James & Johnson.

The present-day distribution of animal phyla in the three major types of environment – the sea, fresh water and land – was described in Section 4.2. From Table 4.1 it is possible to gain some idea of the differences between the composition of the animal communities in these three environments. In this chapter the communities on land will be considered in more detail, and an account will be given of the success of the various taxa in different habitats. The interactions between taxa will then be discussed to provide a brief introduction to the ways in which terrestrial ecosystems differ from aquatic ones.

### 10.1 Composition of animal communities on land

There are many difficulties in establishing to what degree various animal taxa contribute to the variety of terrestrial ecosystems. Not least of these are enormous variations within habitat types, and this problem becomes worse the broader the habitat categories considered. Since the intent of this chapter is to present a broad view, however, regional habitats or biomes will be compared. In an attempt to lessen some of the problems, we can begin by taking one sub-set of the fauna – the soil fauna. This comprises cryptozoic, 'aquatic' and hygrophilic animals (see Table 2.2). Three possible ways of assessing the make-up of these soil communities are taken in turn. The account given is heavily dependent upon the review by Petersen & Luxton (1982), which summarized the work on soil faunas

247

organized by the International Biological Programme, and which should be consulted for further detail.

### 10.1.1    *Abundance of different taxa of soil animals*

In Fig. 10.1 are presented data for abundance (as numbers $.m^{-2}$) of various animal taxa in six biomes. These figures should be taken only as rough guides, and a more complete summary and discussion are given by Petersen & Luxton (1982).

On the criterion of abundance, the smaller animals tend to dominate. Nematodes, in particular, are shown as the most abundant invertebrates in *all* biomes, although still smaller organisms such as protists and bacteria are more abundant again by many orders of magnitude (Alexander, 1977). Testate amoebae, for instance, have been reported at densities of $10^9.m^{-2}$ in woodland soil. Nematodes exist primarily as aquatic animals in the water films between soil particles, but their ability to withstand desiccation in the cryptobiotic state has allowed them to invade even the hot deserts, and even there densities up to $10^6.m^{-2}$ are reached. Rotifers and tardigrades are also active only in water films, but they can enter cryptobiosis during periods of desiccation, and are found in all biomes except hot deserts.

The invertebrates of larger size show a great diversity in most biomes. The oligochaetes are widespread from arctic tundra to tropical rain forest, but although found in dry tropical soils they have not penetrated into hot deserts. The gastropods, in contrast, have failed to reach the high latitudes, but because they are able to aestivate they are not infrequent in hot deserts. They are more common in deciduous than coniferous forests. In deciduous litter there are many shelled species, but under conifers these are rare, and slugs predominate.

Of the crustaceans, perhaps the most surprising are the copepods. They are essentially aquatic, like nematodes, but must also be able to enter cryptobiosis, since harpacticoid copepods are present even in the arid desert soils of Australia, together with some water fleas (cladocerans) (Wood, 1971). The amphipods are restricted to humid environments, and are also constrained mainly to the southern hemisphere: although common in south temperate forests, they are rare in the north temperate equivalents, except in Japan. This presumably relates to problems of distribution, and not to environmental tolerance. The isopods are more widespread, reaching north temperate zones and even forming substantial populations in hot deserts. Here they exist by using burrows as refuges.

Most of the high arachnid numbers are due to enormous densities of mites. The densities are generally higher in wooded than in non-wooded areas, and within grasslands numbers diminish from temperate zones to the arctic tundra. In hot deserts, also, numbers are low. Within the wooded areas, numbers are generally higher in temperate deciduous and northern coniferous zones than in tropical rain forest. Overall, densities appear highest in the temperate zone, but the reasons for this are not well understood. Available moisture, soil structure, temperature, organic content and pH have all been offered as controlling factors.

The apterygote insects in soil comprise mostly collembolans, which overall show rather similar distributions to the mites. There is, in this case, no real difference between densities in wooded and non-wooded areas, but densities are higher in temperate, north-temperate and even tundra zones than in tropical regions. As with the mites, a number of explanations for these distributions have been proffered, including soil moisture, pH and organic content.

Larger arthropods such as myriapods are more restricted in their distribution and maximum densities. They are absent from tundra and hot deserts, but common in all forest types and temperate grassland. Pterygote insects, often represented by larval forms, are abundant in all biomes, although composition of their communities varies enormously. In tundra habitats, for instance, dipteran larvae make up the bulk of the insect numbers, whereas ants and termites account for high proportions of the fauna in the tropics.

Overall, the greatest diversity of groups can be seen to occur in temperate forests, temperate grasslands and tropical rain forest. Absolute densities of animals, on the other hand, tend to be low in tropical forests, and this point will be discussed later. Diversity but *not* density is low in tundra, while both diversity *and* density are low in hot deserts. Some discussion of the factors related to the differing levels of diversity will be given in Section 10.1.3. For the present, we need to investigate further the factors relating to overall abundance.

In temperate forests and grasslands, densities of $10^5 . m^{-2}$ are normal for many invertebrate taxa. In the northern tundra, densities may reach this level, but $10^4 . m^{-2}$ would be more characteristic. In tropical rain forest typical abundances are lower still, often as low as $10^3 . m^{-2}$, and in hot deserts $10^3 . m^{-2}$ is the *maximum* reached. We are far from understanding this distribution of abundance, but probably the most common tendency is to attribute diminution in abundance to harshness of physical conditions.

For hot deserts this explanation may be amplified by reference to the very limited rainfall and the very high temperatures, both of which are

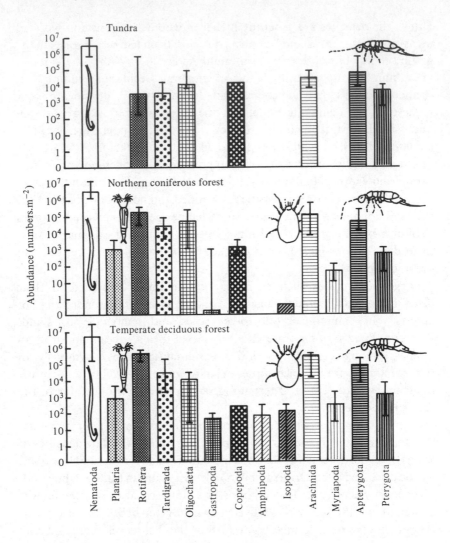

Fig. 10.1. Abundance of the major soil invertebrates in six selected terrestrial biomes. Histograms show average values, and bars show ranges. Values are both varied and approximate, and should only be taken as giving an indication of absolute abundance. After Little (1983), modified particularly by reference to Petersen & Luxton (1982).

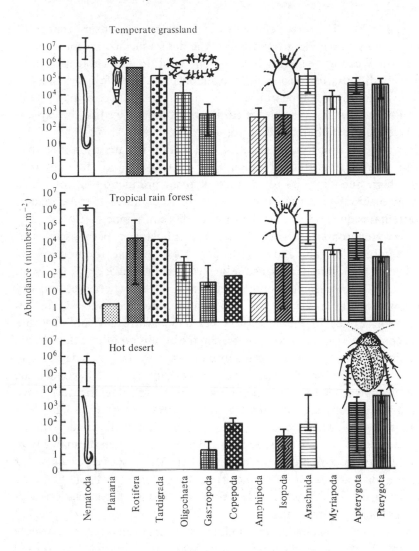

inimical to life. The direct effect of these factors is to ensure that most invertebrates stay underground below the desert surface except at night time. Many can aestivate for long periods until the rains come. In these harsh conditions, there is also little available food, and life cycles must be geared to its seasonal appearance. This physical dominance of the system, with little opportunity for biological interaction, may well be the explanation for reduced abundance of populations.

For tropical rain forests, such an explanation is patently too simple. These environments are well known for their relatively constant environment, high humidity and abundance of plant production, all of which should ensure high densities of soil invertebrates. In fact, some macro-invertebrates do show similar densities to those of temperate forests, but the micro-arthropods such as mites and collembolans are perhaps an order of magnitude less abundant. Several factors combine to produce the relatively small abundances actually recorded (Wallwork, 1976). First, there usually *is* seasonal change in the rain forest, where wet and dry seasons alternate. In the dry season, leaf litter is not decomposed, and lies on the soil surface. In the wet season, high temperature and rainfall ensure ideal conditions for the decomposers, and the litter is rapidly broken down. Leaf litter is therefore incorporated into the soil only to a small degree, and the soil itself therefore often has a very low organic content. Secondly, the structure of tropical forest is rather different from that of temperate forests, in that the canopy usually provides a very effective barrier to light, and there is no well-developed layer of vegetation below it. Together with the development of a varied epiphytic flora, this has meant that the majority of invertebrates are found *above* ground level. Many species of ants, for example, live arboreally in tropical rain forest, and much of the fauna characteristic of the soil in other forests is found in association with arboreal epiphytic bromeliads.

In summary, the abundance of soil invertebrates in all biomes is strongly governed by vegetational characteristics. Where these provide a good food supply and moderate micro-climates, abundances are high. Where they are absent, as in hot deserts, or far removed from the soil surface, as in tropical forest, abundances are low. As we shall see later, these relations may also apply to biomass, but not necessarily to species diversity.

### 10.1.2   *Biomass and energy flow in soil and litter faunas*

When the abundance of various taxa is compared with the picture for biomass distribution, a quite different situation emerges (Fig. 10.2). Soil communities no longer show comparable proportions made up of many taxa. Instead, each biome is dominated by a few animal types. In the main, the oligochaetes are the overwhelming feature of all soils except those in hot deserts. Earthworms in particular are very important in forming the soil structure, and in incorporating surface plant debris into deeper levels, and their distribution has a major influence on energy flow in soil communities. In the litter overlying soils, several other groups are important in recycling detritus. These include the amphipods, isopods, myriapods and gastropods, and in some areas these taxa show high levels of biomass. In northern coniferous forests, yet another group dominates: these are the mites, which can equal in biomass the enchytraeid worms. In general though, the animals second in importance to the oligochaetes in terms of biomass are the pterygote insects. In temperate latitudes these are mainly larvae, but in hot deserts the biomass is made up mainly of adults of species such as burrowing cockroaches and tenebrionid beetles.

Two features of the distribution of soil-fauna biomass have been discussed by Petersen & Luxton (1982), both of which relate in some degree to the distribution of oligochaetes. First, there is a difference between mull soils, in which organic material is rapidly decomposed and well mixed into the mineral fragments, and mor soils, in which litter accumulates in a distinct layer above the mineral portion. Mull soils have a high soil-fauna biomass, and are dominated by the Lumbricidae or earthworms. These soils are characteristic of deciduous forest and temperate grassland. Mor soils have a lower total soil-fauna biomass, and are dominated by the Enchytraeidae, together with significant proportions of mites. These soils are characteristic of coniferous forests. The importance in such forests of the mites is not really evident from Fig. 10.2 because the dominance of oligochaetes appears overwhelming. This figure shows biomass as wet weight, and since oligochaetes have a very much higher water content than mites (86% as compared with 59% respectively), their dominance is emphasized. Dry weight estimates would emphasize the importance of mites.

The second feature of soil biomass to be emphasized by Petersen & Luxton (1982) concerns latitudinal gradients, which are to some extent related to the distribution of mull and mor soils, but also to temperature. Maximum overall biomasses of forest soil faunas have been recorded in temperate and subtropical regions. Both northern coniferous forest and tropical rain forest have lower totals. These overall totals hide the fact that larger animals – the macrofauna – such as earthworms, millipedes and molluscs, have their greatest biomass in temperate and subtropical forests, while smaller animals – the mesofauna – such as mites and enchytraeids,

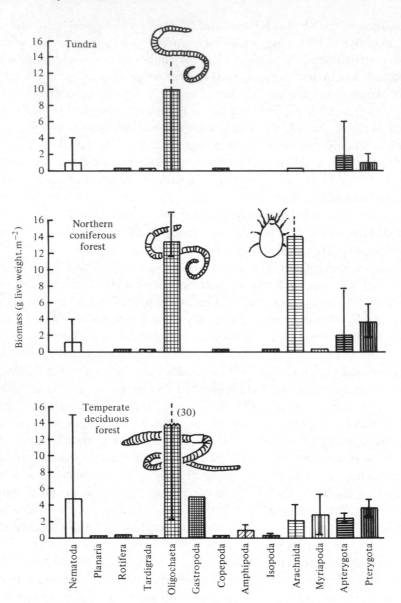

Fig. 10.2. Biomass of the major soil invertebrates in six selected terrestrial biomes. Histograms show average values, and bars show ranges. Values are both varied and approximate, and like the figures for abundance in Fig. 10.1 should only be taken as giving approximate indications. After Little (1983), modified particularly by reference to Petersen & Luxton (1982).

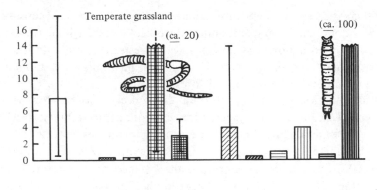

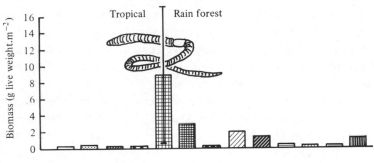

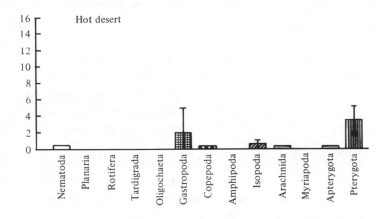

have their greatest biomass in northern coniferous forests. Macrofauna therefore dominate in temperate and subtropical forests, while mesofauna dominate in both northern and tropical forests. A similar latitudinal picture is found in grassland systems, and these have soil-fauna biomasses roughly comparable with those in forests.

The explanation for these latitudinal differences is not at all obvious. Petersen & Luxton (1982) suggested that environmental stability, in terms of drought, waterlogging, frost and extreme temperatures, may be a major controlling factor. For a comparison between temperate zones and tundra this seems a reasonable assumption: the balance of precipitation and evaporation changes drastically between these zones, and soil fauna inhabiting arctic conditions must withstand immense seasonal fluctuations in climatic regimes. Biomass in the tundra may therefore be physically constrained. In tropical rain forest, however, conditions are usually much more constant as discussed in Section 10.1.1, and the low biomass levels there must be determined by other factors.

The importance of the soil fauna in decomposition and nutrient circulation is dependent to a great degree upon its productivity or turnover rate, and hence only indirectly upon its biomass. One approach is to estimate total respiratory metabolism of the soil fauna, but since respiratory rates vary widely with both size and taxonomic grouping, this assessment is quite complex. Figures given by Petersen & Luxton (1982) vary from 24 to 3464 $KJ.m^{-2}.yr^{-1}$. Lowest figures are found in northern coniferous forest (24–662), tundra (30–1799) and tropical rain forest (139–1552). Highest figures are found in temperate deciduous forest (389–2968) and temperate grassland (1961–3464). In parallel with figures for biomass, the temperate zones present the highest respiratory activity. In general, the majority of this is due to the mesofauna, but in grasslands the high densities of earthworms and insect larvae raise the contribution of the macrofauna to equal that of the mesofauna.

To summarize, activity in deciduous forest soils is dominated by earthworms, enchytraeids, nematodes and mites. In coniferous forests, enchytraeids dominate, but mites and collembolans are also important. In temperate grasslands, earthworms, enchytraeids and dipteran larvae are dominant, together with millipedes and gastropods, and amphipods in the southern hemisphere. Tundra soils have no earthworms, but enchytraeids and dipteran larvae dominate, and collembolans are also important in metabolism. This summary should also include the protozoans, which are probably often important in soil metabolism, but whose effects are seldom monitored.

A final point to be raised is the importance of the soil-fauna metabolism when compared with total soil metabolism. In most soils, the fauna accounts for only about 10% of the total respiratory metabolism, the remaining 90% being attributable to the microflora – bacteria and fungi. Nevertheless, the fauna plays a major role in structuring the soil and recycling nutrients because it fragments organic matter and makes nutrients and organics available to the microflora. It has been estimated from exclusion experiments that the fauna processes up to 40% of the annual input of dead organic matter, and in a number of studies it reduced the standing crop of litter on average by 23% per year. The influence of the soil fauna on its own habitat is therefore immense.

### 10.1.3 Species numbers in soil and litter faunas

Although many soil taxa penetrate into the high latitudes, species numbers within any taxon generally fall towards the poles. This is well exemplified by the nematodes and enchytraeids (Petersen & Luxton, 1982). In Antarctic moss mats, 1–16 nematode species are found, compared with 75–92 in temperate forest soils. For Collembola, Antarctic figures are 1–3 species, while in temperate forests 32 (coniferous) and 60 (deciduous) species have been recorded. A similar reduction is found in hot deserts, and, perhaps surprisingly, in tropical rain forests. The factors determining such differences in species diversity of soil fauna between biomes are probably very complex, and cannot as yet be resolved in detail. Wallwork (1976) pointed out that we do not as yet know if soil populations are in general limited by food supply, or whether some other factor such as pressure of predation or climate keeps them at the observed levels. He suggested that possible ways in which faunal diversity may be related to the type of habitat seem most likely to be related to the diversity of *micro*-habitats. In particular these are determined by mosaics of three factors – micro-climate, physical structure and biochemical make-up – which we can briefly take in turn, using as examples the diversity available in grassland habitats.

Micro-climatic diversity is found in both horizontal and vertical gradients, both these in turn relating to vegetation type and degree of vegetation cover. For instance, one of the effects of tussock formations in grassland is to provide a shaded micro-habitat, often with temperatures 10°C lower than those in surrounding open areas (Delany, 1953). Many arthropods, such as spiders, appear to be distributed according to the mosaic formed by these tussocks, and it is possible that temperature is a major factor here. This suggestion is, however, complicated by the fact that the *structure* of the tussocks is also different from neighbouring regions.

Since the tussocks form a more complex mixture of micro-habitats than the non-tussock areas, it could be that greater animal diversity within them is related to this greater spatial complexity rather than to the more equable temperature regime.

The importance of physical structure is further exemplified by the effects of grazing by large herbivores. Morris (1968) compared the invertebrate fauna of chalk grassland from grazed and ungrazed sites (Fig. 10.3). The grazed areas contained less than half as many individuals as the ungrazed areas, and Morris attributed this difference to reduced amounts of food supply, more variable and extreme micro-climates, and less structural diversity of micro-habitats. In terms of structure, the plants in grazed areas were characterized essentially by roots, leaves and vegetative buds only, while in ungrazed areas the plants had long stems, reproductive buds, flowers, seeds and so on, providing a much greater diversity of available niches. Even this quite complex explanation is certainly too simple, however, since similar work on sandy grassland by Southwood & Van Emden (1967) showed the converse phenomenon of a richer fauna on cut grassland than on uncut sites. The difference between the two sets of results may relate to the height at which the grass was cut or grazed (Wallwork, 1976). In Southwood & Van Emden's work, the grass was not cut close to the ground, and in this case sufficient above-ground structure remained to allow many spiders to remain and build webs. In the work of Morris, grazing occurred close to the ground. If this is the correct explanation, grazing can be seen to be a factor affecting faunal diversity depending upon its intensity. Intermediate grazing pressure causes increased diversity, but maximal and minimal grazing relate to decreased diversity. The importance of such non-linear density-dependent effects has probably been under-rated until recently, although Underwood, Denley & Moran (1983) have emphasized similar points in marine intertidal systems.

The effects of biochemical factors in structuring soil communities must be related both directly to the composition of plant communities, and indirectly to the physical and chemical factors that control plant decomposition and the decomposers. In particular, a major influence is the carbon:nitrogen ratio of the dominant plants (Wallwork, 1976). In forest environments, this ratio tends to be low in acid coniferous litter, but high under deciduous trees. The differences in carbon:nitrogen ratio are correlated with differences in abundance of various types of invertebrates. For instance, nitrogen-poor coniferous forests have a wide diversity of micro-arthropods, but not of macrofauna. Nitrogen-rich litter, on the other hand, has a much wider diversity of earthworms, millipedes, woodlice and snails.

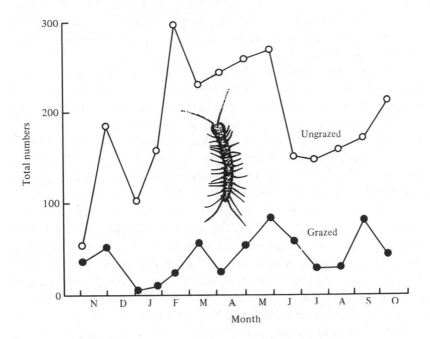

Fig. 10.3. Numbers of lithobiomorph centipedes in grazed and ungrazed chalk grassland over a period of one year. Numbers refer to the totals extracted from four turves of approximate total area 0.3 m². After Morris (1968).

With these three factors in mind, it is possible to return to the individual biomes, and to make some comparisons between them. Comparing the temperate forests, where diversity of soil fauna is high, with tundra and hot desert, where it is low, it is likely that micro-climatic, structural and biochemical diversities are *all* lower in the extreme climates than in the temperate zone. For example, the range of plant species is much restricted in both tundra and hot deserts, and this will lead to a low degree of variation in structural niches and in the composition of the leaf litter. In addition, the variations in micro-climate found in tundra and hot desert may well be over-ridden in importance by the harshness of the macro-climates there. Extremely low temperatures in the tundra, and high temperatures in the desert, may be the major factors restricting species diversity. As discussed by Usher (1985), harsh climates may promote 'adversity selection'. Species are selected for their tolerance of physical conditions, and these conditions keep diversity so low that there are few interactions between species. It is also worth noting that this selection may not be solely for *tolerance* of individuals, but for ability of the individuals and their descendants to tolerate physical extremes and *then* to reproduce rapidly after catastrophic mortality. Such adversity selection would be effective in harsh conditions where the environment is fierce but predictable – the case in tundra and desert.

This leaves for discussion the case of the tropical forests, already considered briefly in Section 10.1.1. Here the physical conditions are often thought to be very favourable and constant. In fact, however, such generalizations are not true for many soils in tropical rain forest. Seasonality has already been discussed in Section 10.1.1. The very high rainfall is also important because it produces rapid leaching of nutrients from the surface layers, and often causes waterlogging. In these circum-stances the soils may become anoxic. Since the organic detritus rapidly decays in the wet season, there is also a reduced food supply for the detritivores. Added to these points is the emphasis by Wallwork (1976) that much of the rain-forest fauna is found *above* the soil. Even in the litter zone, conditions are more favourable than in the soil, and animal diversity there may be high. Williams (1941) recorded more than 200 species of invertebrates from leaf litter in Panamanian rain forest. In this micro-habitat insect and mite species numbers were high, as in temperate rain forests, but many other taxa were also abundant. Among these were soft-bodied animals such as leeches, flatworms and molluscs. This high diversity may reflect the structural and biochemical diversity of micro-habitats, determined in turn by the high diversity of tree species. Individual tree species in tropical forest seldom form pure stands like those in temperate forests, and their root systems, as well as their various leaf and seed types, form a complex mosaic on the forest floor.

This consideration of species diversity in regions above the soil itself leads on to a discussion of species numbers in terrestrial environments as a whole, and the factors determining this diversity.

### 10.1.4    *Species numbers in the overall land habitat*

An estimate of the numbers of species of organisms extant in the world at the present day was made by Southwood (1978). The plants (excuding algae) total about 300 000. Animals total about 1 million, and of these nearly 800 000 are insects, while only 54 000 are vertebrates. The dominance of the insects is even more striking when estimates are restricted to terrestrial animals (Table 10.1). It should also be noted that the figures given for insects are conservative. Matthews (1976) has predicted that in fact there may be 4 to 10 million insect species. With the arachnids also, the true total is unknown. There are 32 000 species of spiders and 25 000 species of mites known, but there are undoubtedly many more species of mites as

Table 10.1 *Approximate numbers of terrestrial species in some animal taxa*

| Taxon | Number of terrestrial species |
|---|---|
| Tricladida (flatworms) | 500 |
| Nemertea (nemertines) | 20 |
| Gastropoda (gastropod molluscs) | |
|    Prosobranchia (operculates) | 4000 |
|    Pulmonata (stylommatophorans) | 20 500 |
| Oligochaeta (earthworms and enchytraeids) | 2000 |
| Onychophora (e.g. *Peripatus*) | 70 |
| Isopoda (woodlice) | 1000 |
| Amphipoda (landhoppers) | 50 |
| Arachnida (scorpions, spiders, etc.) | > 64 000 |
| Myriapoda (centipedes, millipedes, etc.) | 11 000 |
| Hexapoda (insects) | >750 000 |
| Vertebrata | |
|    Amphibia (frogs, salamanders, etc.) | 2000 |
|    Reptilia (snakes, lizards, etc.) | 5000 |
|    Aves (birds) | 9000 |
| Mammalia (mammals) | 4000 |

*Note:* Numbers have been rounded up and are only approximate.

*Source:* After Little (1983); data from many authors.

yet undescribed. Third in order of numbers are the gastropods and just below them the vertebrates. The oligochaetes, although comprising a large biomass, have only 2000 species, of which 1300 belong to one tropical earthworm family, the Megascolecidae. Other groups have even smaller numbers.

The reasons for this varying species richness on land may hold the key to understanding much of terrestrial ecology. Yet no overall explanation is readily available. Success on land may relate in some degree to physiological tolerance, which must set the number of types of micro-habitat that can be exploited; but such an explanation really only pushes this question further back: why are some species more tolerant than others? Some taxa, moreover, appear to have not only greater tolerance limits, but more ability to take advantage of different lifestyles. The flatworms, nemertines, onychophorans, and amphibians, for instance, have remained primarily predatory. The gastropods, oligochaetes, isopods, amphipods and myriapods have remained mainly detritivores and herbivores. In contrast, the insects, arachnids, birds and mammals have radiated widely in nutritional terms and show predatory species, detritivores and herbivores. In particular, the insects have exploited every nutritional niche, including those of parasite and symbiont.

Some of the reasons why different animal taxa contain widely different numbers of species have been discussed by May (1978). Reviewing influences on insect species numbers, he concluded that the main reason why there are so many species of insects compared with other animals is to do with their relatively small size. From crude estimates of the size ranges of *all* terrestrial animals, he showed that there was a relation between size and species numbers of approximately $S\alpha L^{-2}$, where $S$ is number of species and $L$ is length. This relation is shown in Fig. 10.4, from which it can be seen that there is a rough correlation at sizes above 10 mm. It was argued by May that size classes below 10 mm may be under-represented because of the lack of taxonomic interest in smaller species such as mites. May also considered the possibility that for size classes below 1 mm, particularly for bacteria and viruses, the species concept might itself be a loose one, and that the decrease in numbers of species at these very small sizes might be more apparent than real. It might also be argued that the numbers of described insect species are higher than for other groups because they have more easily distinguishable characteristics. The correlation between size and species number admittedly, therefore, involves a number of approximations and assumptions. Nevertheless, whatever the reason for it, the reality of the relationship $S\alpha L^{-2}$ on present evidence is convincing.

For insects, at least, the factor of small size allows species to occupy smaller niches in the terrestrial world than larger species such as the vertebrates. Thus Southwood (1978) has discussed how small size relates in particular to an organisms's 'trivial range' – the area in which it feeds. Since small animals have, in general, smaller trivial ranges, they experience the environment as more patchy than do larger animals. For example, the trivial range of a 1 mm long mite would be about 80 mm, equivalent to a grass blade. Movement off this blade would take the mite into quite a different micro-habitat. On the other hand, the trivial range of a deer would be of the order of hundreds of meters, and when the deer moves within that range it will still experience only a few micro-habitats. The mite, within the same area, could occupy thousands of trivial ranges.

Small animals like insects and mites have therefore evolved adaptations to the small patches or micro-habitats available, and this may have increased their numbers of species. Lawton (1986) has considered the increase in number of insect species in relation to available surface area on plants in more detail. Smaller insects may have more spaces available to

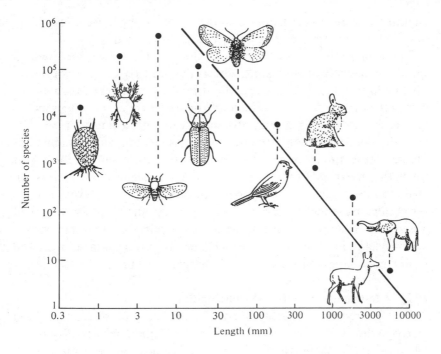

Fig. 10.4. Approximate numbers of species ($S$) of all terrestrial animals, plotted in terms of length ($L$). The solid line shows the shape of the relation $S \alpha L^{-2}$. Circles show estimated numbers. Note that protozoans are mostly *below* 0.3 mm in length; the illustration of a testate amoeba is given to emphasize the importance of protists at the small end of the scale. After May (1978).

them because of the nature of plant surfaces, and at least some of the increase in small species is due to this as well as to the increase in variety of spaces available to them. Small species also breed faster than larger ones, and this may lead to higher absolute rates of speciation, reduced rates of extinction, and hence more small species (Lawton, 1986).

Size does not, however, explain the difference in abundance of, say, terrestrial insects and terrestrial crustaceans. Apart from physiological tolerance, mentioned above, an important point here is the ability to disperse. Insect flight mechanisms allow rapid movement to new habitats, and this capacity may be particularly important in exposing insects to new selection pressures. The highly mobile insects may, for instance, land on plants that are not their normal hosts. Although most will immediately depart, some may remain and utilize the new plant for feeding (Southwood, 1978). In this case there is the possibility first of the range of the species being extended; and second of the formation of a new species if the migrants become isolated from the main bulk of the population. The ability to disperse may also therefore contribute to speciation.

Southwood (1978) has also pointed out another reason for the high species numbers found in insects. This is the property of metamorphosis, which allows many insects to inhabit at least two different environments during their life cycle. This movement from one habitat to another means that each species uses one environment for only a limited time. The possibility of competition with other species is reduced, and in many cases similar species are separated in the resources they use mainly by the time at which they use them. In contrast, the vertebrates and most other terrestrial invertebrates use similar environments throughout their life cycle. For the vertebrates in particular, life cycles are too long to allow any subdivision of a particular resource between juveniles and adults on a seasonal basis.

Consideration of the ways in which various terrestrial taxa interact to form communities must now be taken in more detail.

## 10.2    Interactions: terrestrial communities

Terrestrial foodwebs are based on primary production by green macrophytes, and in most parts of the world the angiosperms are the dominant macrophyte group. Only in high latitudes do the gymnosperms predominate, where they form coniferous forests. Much of the diversity of terrestrial communities can therefore be related, directly or indirectly, to angiosperm diversity. However, the *effective* diversity of angiosperm material varies immensely depending upon the mechanism by which it is degraded. In the next two sections the effects of direct herbivory will be discussed. In the third section the effects of herbivores will be compared with those of detritivores.

### 10.2.1    Herbivore-plant relations

Plant material may be consumed directly by herbivores, which themselves show many different mechanisms of attack, and many specific feeding adaptations related to leaf shape, structure and chemical composition. Their grazing has in turn induced defensive mechanisms by the plants, so that grazer-plant interactions form a complex determinant of terrestrial community structure. While these interactions concern all herbivores, the subject has been most fully explored for insects (Southwood, 1986). Insect species are usually specialized for feeding on a few plant species, and often only on one restricted part such as roots, stem, leaves or flowers. Within each of these subdivisions, there may be several different mechanisms of attack. Root (1973) defined, for a cabbage plant, three separate guilds of leaf-feeding insect herbivores: a strip-feeding guild, a pit-feeding guild, and a sap-feeding guild. It might be thought that for these leaf-feeders the food supply would be easy to obtain, but in fact there are many points that make feeding on angiosperms more difficult than it seems (Edwards & Wratten, 1980).

First, the micro-climate surrounding the leaves is often more severe than that on the ground. As discussed by Willmer (1986), humidity may be high close to leaf surfaces, but the boundary layer is often very thin, and the upper surfaces of leaves may have much higher temperatures than surrounding air. Feeding on plants may therefore expose insects to problems of desiccation.

Secondly, for animals the size of insects, attachment to the plant and access to the leaf tissue may not be straightforward. Many leaves are covered by trichomes – plant hairs – which make it difficult for insects to move about on the leaf surface, and to grasp the leaf proper. It is true that some species can actually use the trichomes to hold on to, but in general they are seen as defensive structures evolved by the plants (Southwood, 1986). Some of these trichomes may be glandular, and secrete sticky or toxic substances which further deter insect herbivores.

Thirdly, most plants produce chemical substances in their tissues that have no direct role in normal metabolism, but act to impair the health of the herbivores. These 'secondary plant substances' have apparently evolved in relation to insect herbivory, and in turn many insects have evolved responses allowing them to avoid, tolerate, or even make use of, the effects of specific chemicals. Inevitably this has led to specialization by both plants and insects, so that at the present day the majority of insect herbivores are quite selective about their plant hosts, and the majority of plant species are protected against a variety of potential pests. This 'evolutionary arms race' will be considered in the next section under the term co-evolution.

### 10.2.2 Insect and vertebrate herbivores

The relations between insects and plants, and between vertebrates and plants, are quite different. Since terrestrial communities today are dominated by insects, vertebrates and angiosperms, some discussion of how they all interact is essential for the understanding of the terrestrial ecosystem as a whole.

Insects, as described above, are small enough to experience a diversity of micro-habitats within and around green plants. It is probable that their association with angiosperms began as one in which they consumed plant detritus rather than the living tissue (Hughes, 1976). After that, however, direct herbivory arose to some extent, and in spite of the protective systems of plants, many insect adults and larvae feed directly on plant tissues. In some cases very specific interactions between insect and plant species have evolved. Here we shall consider just two examples.

The first of these concerns ants, which may have very complex relationships with plants, but in general do not eat plant tissues directly (Huxley, 1986). Several species of ants, however, act as defenders of plants against potential herbivores. They disturb visiting insects before they have time to steal nectar or eat tissues. The ants are attracted in the first place by nectaries not involved in the flowers themselves (extra-floral nectaries). They use these as a food source, and defend them – and the rest of the plant – against invaders. Some plants also produce chambers which are used by the ants as refuges ('domatia'), and sometimes nectar is provided in these domatia as a food supply. Alternatively, sap-sucking insects may colonize the domatia, and the ants obtain honeydew from them by tending and milking them. In most of these ant-plant relationships, the adaptations seem primarily to concern the plants. The ants, in contrast, show few permanent modifications, and indeed have no obligate relationships with the plant. In general, then, these relationships do not involve what is called 'co-evolution' – reciprocal evolutionary changes in *both* interacting species. Nevertheless, the influence of the ants in determining species composition of insect communities on plants can be very great.

The second example concerns butterflies of the genus *Heliconius* and two species of vines (Gilbert, 1982; Strong, Lawton & Southwood, 1984). *Heliconius* spp. live in the tropical forests of Central and South America. Their larvae feed on vines of the genus *Passiflora* (passion flowers). This relationship is very species-specific, because the *Passiflora* vines produce a variety of toxic chemicals such as glycosides and alkaloids, and each species of *Heliconius* is adapted to tolerate only a restricted combination of these toxins. The adult butterflies feed on the nectar of another vine genus, *Anguria*, which rely on them for pollination, and which provide nectar with a high nutrient content. This allows the adults to live for up to 6 months, during which time they visit the very scattered plants of *Anguria* thus ensuring cross-pollination. When the butterflies come to lay their eggs on the *Passiflora* vines, several mechanisms evolved by the plants act as deterrents. In particular, some species have produced structures which closely resemble *Heliconius* eggs. Since adults usually lay eggs only on shoots without eggs already present, these mimic-eggs deter egg-laying. The complex interactions between *Heliconius, Anguria* and *Passiflora* have been interpreted by Gilbert (1982) as a series of stepwise co-evolved responses (Fig. 10.5).

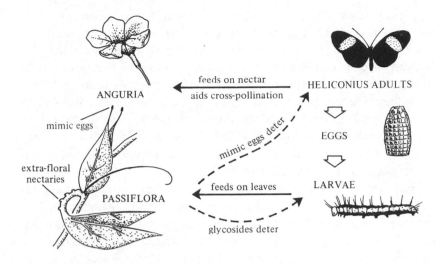

Fig. 10.5. Selected interactions between butterflies of the genus *Heliconius* and the two vines, *Anguria* and *Passiflora* in Central and South American rain forest. Solid arrows show trophic interactions, dotted lines show deterrents, and open arrows show life cycle. From information of Gilbert (1982).

A major question arising from the work on *Heliconius* is whether such co-evolution is common in insect-plant relationships. If it *is* common, it might be fair to say that the present diversity of both angiosperms and insects is in large measure due to such reciprocal evolutionary steps. In an analysis of this question, however, Strong, Lawton & Southwood (1984) concluded that such 'special and mutual counteradaptations' are not typical of insect-plant relationships. Instead, they considered that the majority of relationships could be explained in terms of 'diffuse coevolution' (Janzen, 1980). This process is a rather less specific one, in which plants evolve in relation to several insect species, and insect adaptations relate to more than one plant species. In its widest terms, such a process might indeed explain the rise in diversity of both plants and angiosperms, but without the closely circumscribed links found in the *Heliconius – Anguria – Passiflora* triangle.

Vertebrate herbivores contrast strongly with the insects in their relationships with the angiosperms. In particular this is because they are mostly much larger than insects, and therefore do not experience the diversity of niches presented by angiosperms to smaller animals. When grazing, for instance, many vertebrates eat whole plants or large parts of them, and do not choose between specific groups of cells like the insects. There are thus very many fewer species of vertebrate herbivores than of insects. This difference is probably enhanced because of the difference in generation times: in vertebrates these are usually long, and perhaps in association with this the phenomenon of learning has evolved to allow them to react to environmental change. In the insects, generation times are short, and there is insufficient time for the evolution of learning to have been prompted by environmental fluctuations (Slobodkin & Rapoport, 1974). Instead of using learning, the insect genotype has responded by rapidly developing new species to adapt to changing conditions. Consequently, the number of insect species at any one time can be enormous.

There are, however, some interesting interactions between vertebrates and angiosperms, and some of these also relate to insect communities. For instance, Southwood (1986) has discussed the plants that produce stinging hairs. These hairs seem to have evolved as a defence against vertebrates, and not against insects. Plants such as the nettle, *Urtica dioica*, and several others in the same family, such as the Australian forest tree *Laportea*, are avoided by mammalian herbivores. In tests using rabbits and sheep, Pollard & Briggs (1984) showed that nettle plants with reduced densities of stinging hairs were grazed more than those with normal densities. Invertebrates such as slugs and snails were not apparently deterred. Indeed, Southwood (1986) suggested that perhaps because of the decreased damage to these plants by vertebrates, the insect faunas on them have become particularly rich and diverse. It is also possible, but unproven, that birds may avoid feeding on nettles, and this also would allow the insect communities to diversify. In effect, it may be that because the plants have concentrated upon defence against vertebrates, they have become more susceptible to insects.

The more widespread phenomenon of vertebrate grazing on grasses was discussed earlier (10.1.1). Grazing pressure reduces invertebrate density and diversity when extreme, but moderate grazing may promote a more diverse fauna. Vertebrate grazing can therefore influence invertebrate communities, but not because of any type of co-evolution with plants.

In summary, the effects of insect and vertebrate herbivores on community structure are quite different. Insects have speciated widely in relation to angiosperm structure, and plants have evolved many and various defence mechanisms against them. Vertebrates, primarily because of their large size, have not become such specialist feeders, and their effects on community structure are on a much broader scale, in relation to their rather more gross feeding behaviour.

### 10.2.3 Detritivore – plant relations

Although there are more phytophagous *species* of insects than detritivorous species, the direct herbivores are restricted to only 9 orders, compared with a total of 29 orders of living insects (Strong, Lawton & Southwood, 1984). The problems of feeding on green plants, discussed above, evidently do constitute an appreciable deterrent to herbivory. Whether because of this, or not, much material produced by plants passes through a separate food web, the detritivore system, in which detritivores eat the dead or decaying parts of plants, and gain energy at least partly by digesting the saprophytic bacteria and fungi that are involved in decay processes. When vegetable matter is consumed in this form – as detritus – much of its inherent diversity is lost. Most detritivores eat a variety of plant types, which lose much of their variety as leaf shapes and structures are broken down and fragmented. The detritivore lifestyle therefore probably has a unifying, rather than a diversifying influence. To some degree, this difference between herbivores and detritivores is reflected in the species numbers of some groups. Oligochaetes and prosobranch gastropods, for example, both primarily detritus feeders, and important in the turnover of leaf litter, have only 2000 and 4000 species respectively. The amphipods, also detritivores, have only 50 species. In contrast, the pulmonate gastropods, many of which are direct herbivores, have 20 000 species. Within the insects, too, the number of species in decomposer communities is probably lower than that in grazer communities (Usher *et al.*, 1979). Within the decomposer communities, those in woodland are more diverse than those in grassland, which in turn are more species-rich than those in moorland. Whether this variation in diversity can be related directly to the diversity of detrital food, or to quite different features of the communities, is unknown.

### 10.2.4 Predation and its effects on community structure

As pointed out earlier (9.6), there is a large diversity of predatory taxa on land. These include a wide array of types, from the soft-bodied flatworms and nemertines to the arthropod forms such as centipedes. The major predators on land, however, are the arachnids, insects and vertebrates. The present-day spiders are dependent upon the insects for prey, and there is no doubt that the rise of the spiders was consequent upon the adaptive radiation of the insects. The interactions between these two groups now form a major part of terrestrial ecosystems, although the insect predators themselves probably prey primarily on other insects, forming a rather circumscribed community. Once again, the vertebrates differ from the invertebrate predators because of their size. Many are specific in the prey that they take, but many reptiles and predatory birds will eat a wide variety of insects. Larger vertebrates such as the mammals tend to be predators of other mammals, as they require a larger food intake than is available from any invertebrate source.

All these trophic interactions vary widely in importance between the different terrestrial biomes. Nevertheless, the importance of predation as a major force controlling community structure in many ecosystems is widely recognized. For communities of phytophagous insects, for example, Strong, Lawton & Southwood (1984) have concluded that populations are kept at low densities by predation, parasitism and disease – so much so that competition between species is hard to demonstrate. For instance, the winter moth, *Operophtera brumata*, whose caterpillars feed mainly on oak leaves, is preyed upon by a wide variety of species (Fig. 10.6). Pupae are killed by ground-living beetles and small mammals during their over-wintering period. Larvae are parasitized by flies and ichneumonid wasps, and are also attacked by protozoan and virus diseases. As a result of all these combined effects, densities are kept low. For other species of phytophagous insects, predation by birds may be important. Although birds do not usually congregate fast enough to regulate a fast-growing insect population, experiments in which birds were excluded from a forest understorey showed that Lepidoptera were normally reduced by the activities of bird predators. Another example is given by Strong (1984), for the insects living on a tropical monocotyledon, *Heliconia imbricata*. Beetles living in the leaf scrolls of this plant had excess food, and showed no evidence of crowding or competition. They were probably maintained at low densities primarily by the effects of predators and parasitoids.

For the effects of spiders as predators, Wise (1984) has pointed out that not only are spiders ubiquitous in terrestrial ecosystems, but they capture a major fraction of the energy in consumer species. There is a wide variety of feeding guilds, from web-spinners to active hunters. Wise noted that, as in phytophagous insect communities, competition between spiders was hard to demonstrate, and hypothesized that, in general, spider densities are reduced by factors such as dispersal, natural enemies and the effects of the physical environment.

For vertebrate predators, Kitching (1986) has outlined the effects caused by introductions of several species by man. The cane toad, *Bufo marinus*, is now abundant in tropical and subtropical communities world-wide. It has a catholic diet, ranging from beetles to earthworms, centipedes and even

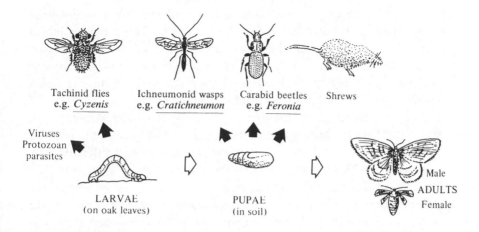

Fig. 10.6. The influence of predation on populations of the winter moth, *Operophtera brumata*. Open arrows indicate the life cycle. Solid arrows indicate significant predators. From information of Varley *et al.* (1973).

small mammals. It is itself toxic to some predators, and has been reported to cause the deaths of several species of reptile that have fed upon it, while some predators have apparently eaten *B. marinus* without harm. In many places it appears to have virtually replaced some species of frogs, skinks and small mammals, but it is unclear whether this has involved direct predation or competition. Certainly, the introduction of this predator has greatly altered community structure.

In summary, the classical view that competition between species was a major driving force structuring communities on land seems now to have been replaced by one in which many species are kept at such low densities by predation that competition does not have a great impact. The evidence for both theories is, however, as yet rather scant, and in much need of further investigation.

### 10.2.5 *The influence of evolutionary time on terrestrial communities*

The trophic interactions discussed in Sections 10.2.1 to 10.2.4 provide some of the major influences in the structure of terrestrial communities. As described in Sections 10.1.1 to 10.1.4, however, community structure varies widely between biomes, and most of these variations are presumably in the final analysis mainly dependent upon climatic and geographical factors. There are, however, further influences upon community composition. One of these, which will not be discussed in detail here, is the biogeographical effect following from the origin of terrestrial groups in particular sites, and their subsequent spread from their points of origin. One example from the soil fauna is provided by the terrestrial amphipods, which are found almost exclusively in the southern hemisphere. This is probably due to the origin of the group in this hemisphere, and its failure to invade regions north of the equator.

Here it is appropriate to consider briefly not so much the *place* of origin of terrestrial groups, but the *time* of their origin. As we have seen in Chapter 1, different groups are thought to have invaded land at very different times. Southwood (1973) has suggested that the longer an insect group has been exposed to particular host plants, the greater is the number of species that utilize the plants. If this principle can be extended to other groups, it might explain why the myriapods have as many as 11 000 species whereas isopods have only 1000: myriapods have probably been on land since the Silurian (over 400 million years), while the terrestrial isopods probably date only from the Permian (less than 300 million years). Terrestrial amphipods, with only 50 species, may have a very recent origin, possibly only a few million years. It may be that, with a more detailed comparison between taxa, further, better justified, correlations would emerge.

## 10.3    Terrestrial and aquatic communities compared

Most marine communities depend for their primary food source on phytoplankton production, or, in the littoral zone, on production by benthic microalgae. On land, the majority of primary production is carried out by large plants, including shrubs and trees. The two systems are therefore quite different in their trophic structure (Odum, 1971; Owen, 1980). In the sea, plant biomass at any one time is very low, but production is high. Phytoplankton production passes to a large degree directly to herbivorous zooplankton, while macroalgal production is little utilized by direct herbivores. Instead, most is degraded to detritus and fuels a large detritivore system. In fresh waters, the system is similar, although macrophyte production there is attributable mainly to angiosperms. The first level of consumers in the sea and fresh waters consists, therefore, mainly of phytoplankton-feeders and detritivores.

On land, plant biomass is large, but a great deal of it is in the form of structural tissue. This tissue, necessary on land to support the photosynthesizing parts of the plants, usually has a very low nutritive value, and even when degraded forms resistant fibrous detritus of little value to detritivores. The green parts of land plants are eaten both directly by herbivores – especially the insects and vertebrates – and also degraded to edible detritus.

In contrast to the situation in the sea, direct grazing on macrophytes is an important process, and grazing has a major influence on terrestrial plant communities. In part, at least, this is due to the differences between the structure of aquatic algae and angiosperms. Plants such as the terrestrial grasses, for example, can withstand grazing to the ground level, because they have roots and buds below ground that are not damaged by grazers. Aquatic macroalgae, however, have no roots, and no below-ground portion from which they can re-grow once grazed. Grazing on aquatic macrophytes would therefore in the long run be an unstable evolutionary strategy. Most aquatic grazers feed instead on microalgae and the sporelings of macrophytes.

The terrestrial grazers and detritivores have to be mobile, in order to seek a continuing food source. There are relatively few static animals on land, unlike the situation under water, where many different kinds of invertebrate are sessile. There are, however, two parallels on land. One is provided by some of the sap-sucking insects, which are tapping such a large resource that they never need to move. These fall an easy prey even to slow insect predators. The second is provided by the web-spinning spiders, which effectively trap aerial plankton. There is, however, no such great *variety* of slow-moving predators as in the sea, where starfish, gastropods or crabs may take many hours or days over a meal, and sessile filter feeders such as barnacles and sponges are abundant. The lack of floating aerial food, compared with the vast diversity of plankton in the sea, therefore accounts for many of the differences in taxonomic diversity between the two environments.

The relations between plants, grazers and detritivores differ in a further way between aquatic and terrestrial ecosystems, as discussed by Remmert (1980). On land, the leaf-fall from plants fuels a detritus food web. The detritivores eat dead leaves, and in so doing break them up into smaller fragments of detritus. This allows more surfaces for the microbes to act upon, and the actions of detritivores and microbes together release nutrients which are then recycled by the plants through their root systems. Without this recycling, which is essentially a property of the soil ecosystem, the nutrient supply would be locked up for long periods in the dead leaves. Growth and regeneration of plants, dependent upon nutrient supply from the soil, would become impossible. The decomposer system of terrestrial soils can therefore be seen to be an essential part of terrestrial ecosystems. The situation in aquatic systems is quite different. Here there is not usually a leaf-fall on to a specific area of substrate, and no real equivalent to a terrestrial soil is formed. Instead, organic matter is quickly broken down and the nutrients are re-utilized or re-dissolved in the water. Since aquatic algae do not have root systems, they do not, in any case, take up nutrients from the substrate but from the water. Although the decomposer organisms are therefore just as important in water as on land, the route followed by nutrient recycling depends upon its distribution in the water mass rather than re-absorption from the soil.

Predators on land also have to deal with a quite different series of problems from those of aquatic predators. In general they operate in two dimensions, while aquatic predators operate in three dimensions (Owen, 1980). As far as the invasion of land is concerned, however, most invaders are pre-adapted to life in two dimensions because they have moved on to the two-dimensional surface of the littoral zone before invading the land surface. The use of three dimensions by birds, bats and insects is a secondary re-discovery, and even then there are many differences from a swimming lifestyle. In particular, most aerial organisms spend only limited times off the ground, whereas aquatic swimmers may spend the majority of time without contacting a surface. This may have much to do with the impossibility of maintaining neutral buoyancy in air, as well as with the scarcity of suspended aerial food.

Terrestrial ecosystems show much greater spatial diversity than aquatic systems, both because physically imposed climates change more suddenly over short distances, but also because of the variety of structures evolved by terrestrial macrophytes – especially the gymnosperms and angiosperms. Even the bryophytes and ferns produce their own micro-climates, as emphasized in Chapter 1. The success particularly of insects and tetrapod vertebrates on land is marked by their ability to respond rapidly to such differences in the environment, and to take advantage of them. These two groups have appeared to monopolize animal communities on land, the one because of enormous numbers of individuals, the other because of large individual size. Terrestrial ecosystems thus appear to be dominated by the effects of macrophytes, and their interactions with insects and vertebrates. It could be argued, then, that the invasions of land by insects and vertebrates, each by entirely different routes, have been the most important factors governing present-day terrestrial animal communities. This would, however, be a short-sighted view. Chapters 4 to 8 have reviewed the origins of many other terrestrial groups, and in the present chapter the importance of a variety of taxa in terms of abundance, biomass and energy flow, have been emphasized. Without the isopods and amphipods, annelids, gastropod molluscs and myriapods, the decomposer systems on land would be remarkably different. Without the arachnids, the predatory systems would also be quite unlike those actually present. Similar remarks apply to the groups of smaller animals such as the nematodes and protozoans, which play major roles in soil ecosystems. Only by considering *all* these groups and their origins can a start be made towards understanding the ways in which terrestrial ecosystems have come into being. It is hoped that this book has made some contribution towards such an understanding.

# References

Adam, P. 1981. The vegetation of British saltmarshes. *New Phytol.* **88**, 143–96.

Agnew, D.J. & Taylor, A.C. 1986. Seasonal and diel variations of some physico-chemical parameters of boulder-shore habitats. *Ophelia* **25**, 83–95.

Ahearn, G.A. 1970. The control of water loss in desert tenebrionid beetles. *J. exp. Biol.* **53**, 573–95.

Alexander, M. 1977. *Introduction to soil microbiology.* 2nd edn. New York: J. Wiley & Sons.

Alvarado, R.H. 1979. Amphibians. In *Comparative physiology of osmoregulation in animals*, ed. G.M. Maloiy, vol. I, pp. 261–303. London: Academic Press.

Andrews, E.B. & Little, C. 1972. Structure and function in the excretory systems of some terrestrial prosobranch snails (Cyclophoridae). *J. Zool., Lond.* **168**, 395–422.

  1982. Renal structure and function in relation to habitat in some cyclophorid land snails from Papua New Guinea. *J. moll. Stud.* **48**, 124–43.

Andrews, S.M. 1973. Interrelationships of crossopterygians. In *Interrelationships of fishes*, eds. P.H. Greenwood, R.S. Miles & C. Patterson, pp. 138–77. London: Academic Press.

Appleton, T.C., Newell, P.F. & Machin, J. 1979. Ionic gradients within mantle-collar epithelial cells of the land snail *Otala lactea*. *Cell. Tiss. Res.* **199**, 83–97.

Autrum, H. (ed.). 1979. *Comparative physiology and evolution of vision in invertebrates A: Invertebrate photoreceptors.* Berlin: Springer Verlag.

Avery, R.A. 1979. *Lizards: a study in thermoregulation.* London: Arnold.

Bacon, M.R. 1971. Distribution and ecology of the crabs *Cyclograpsus lavauxi* and *C. insularum* in northern New Zealand. *N.Z. J. mar. f.w. Res.* **5**, 415–26.

Bahl, K.N. 1947. Excretion in the Oligochaeta. *Biol. Rev.* **22**, 109–47.

Bailey, T.G. 1971. Osmotic pressure and pH of slug haemolymph. *Comp. Biochem. Physiol.* **40A**, 83–8.

Ball, I.R. 1981. The phyletic status of the Paludicola. *Hydrobiologia* **84**, 7–12.

Bareth, C. 1968. Biologie sexuelle et formations endocrines de *Campodea remyi* Denis (diploures campodéides). *Rev. Ecol. Biol. Sol.* **5**, 303–426.

Barnes, R.S.K. 1968. Individual variation in osmotic pressure of an ocypodid crab. *Comp. Biochem. Physiol.* **27**, 447–50.

  1980. *Coastal lagoons.* Cambridge University Press.

  1981. An experimental study of the pattern and significance of the climbing behaviour of *Hydrobia ulvae* (Pennant). *J. mar. biol. Ass., U.K.* **61**, 285–99.

  1984. *Estuarine biology.* 2nd edn. London: Arnold.

  1986. Daily activity rhythms in the intertidal gastropod *Hydrobia ulvae* (Pennant). *Est. coast. Shelf Sci.* **22**, 325–34.

Barnes, R.S.K., Williams, A., Little, C. & Dorey, A.E. 1979. An ecological study of the Swanpool, Falmouth. IV: Population fluctuations of some dominant macrofaunas. In *Ecological processes in coastal environments*, eds. R.L. Jefferies & A.J. Davy, pp. 177–97. Oxford: Blackwell.

Bartholomew, G.A. & Heinrich, B. 1978. Endothermy in African dung beetles during flight, ball making and ball rolling. *J. exp. Biol.* **73**, 65–83.

Beadle, L.C. 1957. Respiration in the African swampworm *Alma emini* Mich. *J. exp. Biol.* **34**, 1–10.

Berneys, E.A. & Chapman, R.F. 1974. Changes in haemolymph osmotic pressure in *Locusta migratoria* larvae in relation to feeding. *J. ent.* (A) **48**, 149–55.

Berridge, M.J. 1965. The physiology of excretion in the cotton stainer, *Dysdercus fasciatus* Signoret. I. Anatomy, water excretion and osmoregulation. *J. exp. Biol.* **43**, 511–21.

  1968. Urine formation by the Malpighian tubules of *Calliphora*. *J. exp. Biol.* **48**, 159–74.

Berridge, M.J. & Gupta, B.A. 1967. Fine-structural changes in relation to ion and water transport in the rectal papillae of the blowfly, *Calliphora*. *J. Cell. Sci.* **2**, 89–112.

Berrill, N.J. 1955. *The origin of vertebrates*. Oxford University Press.

Berry, A.J. 1964. Faunal zonation in mangrove swamps. *Bull. natn. Mus. Singapore* **32**, 90–8.

Berry, A.J., Lim, R. & Sasekumar, A. 1973. Reproductive systems and breeding condition in *Nerita birmanica* (Archaeogastropoda: Neritacea) from Malayan mangrove swamps. *J. Zool. (Lond.)* **170**, 189–200.

Beumer, J.P. 1978. Feeding ecology of fishes from a mangrove creek in north Queensland, Australia. *J. Fish Biol.* **12**, 475–90.

Bliss, D.E. 1979. From sea to tree: saga of a land crab. *Am. Zool.* **19**, 385–410.

Bolwig, N. 1953. On the variation of the osmotic pressure of the haemolymph in flies. *S. Afr. ind. Chemist* **7**, 113–5.

Boss, K.J. 1978. On the evolution of gastropods in ancient lakes. In *Pulmonates*, eds. V. Fretter & J. Peake, vol. 2A, pp. 385–428. London: Academic Press.

Bourne, G.R. 1985. The role of profitability in snail kite foraging. *J. Anim. Ecol.* **54**, 697–709.

Bousfield, E.L. 1984. Recent advances in the systematics and biogeography of landhoppers (Amphipoda: Talitridae) of the Indo-Pacific region. In *Biogeography of the tropical Pacific*, eds. F.J. Radovsky, P.H. Raven & S.H. Sohmer, pp. 171–210. Bernice P. Bishop Museum Special Publication, no. 72.

Boutilier, R.G., Randall, D.J., Shelton, G. & Toews, D.P. 1979. Acid base relationships in the blood of the toad, *Bufo marinus*. I. The effects of environmental $CO_2$. *J. exp. Biol.* **84**, 289–302.

Bray, A.A. 1985. The evolution of the terrestrial vertebrates: environmental and physiological considerations. *Phil. Trans. R. Soc. Lond.* **B 309**, 289–322.

Bridges, C.R. 1988. Respiratory adaptations in intertidal fish. *Am Zool.* **28**, 79–96.

Briggs, D.E.G. 1985. Les premiers arthropodes. *La Recherche* **16**, 340–9.

  1987. Scorpions take to the water. *Nature, Lond.* **326**, 645–6.

Briggs, D.E.G. & Clarkson, E.N.K. 1989. Environmental controls on the taphonomy and distribution of Carboniferous malacostracan crustaceans. *Trans. R. Soc. Edinb. – Earth Sciences* **80**, 293–301.

Briggs, D.E.G. & Conway Morris, S.C. 1986. Problematica from the Middle Cambrian Burgess Shale of British Columbia. In *Problematic fossil taxa*, eds. A. Hoffman & M.H. Nitecki, Oxford: Clarendon Press.

Brinkhurst, R.O. 1982. Evolution in the Annelida. *Can. J. Zool.* **60**, 1043–59.

Brinkhurst, R.O. 1984. The position of the Haplotaxidae in the evolution of oligochaete annelids. *Hydrobiologia* **115**, 25–36.

Brocas, J. & Cherruault, Y. 1973. Association des systèmes trachéens et circulatoires dans le transport des gas respiratoires chez l'insecte aerien. *Bio-medical Computing* **4**, 173–204.

Brownell, P. & Farley, R.D. 1979. Orientation to vibrations in sand by the nocturnal scorpion *Paruroctonus mesaensis*: mechanism of target localization. *J. Comp. Physiol.* **131**, 31–8.

Burggren, W.W. & Johansen, K. 1987. Circulation and respiration in lungfishes (Dipnoi). In *The biology and evolution of lungfishes*, eds. W.E. Bemis, W.W. Burggren & N.E. Kemp, pp. 217–36. New York: Alan R. Liss Inc.

Burky, A.J., Pacheco, J. & Pereyra, E. 1972. Temperature, water and respiratory regimes of an amphibious snail, *Pomacea urceus* (Müller), from the Venezuelan savannah. *Biol. Bull. mar. biol. Lab., Woods Hole* **143**, 304–16.

Burrows, M. & Hoyle, G. 1973. The mechanism of rapid running in the ghost crab, *Ocypode ceratophthalma*. *J. exp. Biol.* **58**, 327–49.

Burton, R.F. 1965. Sodium, potassium, and magnesium in the blood of the snail, *Helix pomatia*. *Physiol. Zool.* **38**, 335–42.

    1973. The significance of ionic concentrations in the internal media of animals. *Biol. Rev.* **48**, 195–231.

    1983. Ionic regulation and water balance. In *The Mollusca*, eds. A.S.M. Saleuddin & K.M. Wilbur, vol. 5 part 2, pp. 291–352. New York: Academic Press.

Cameron, J.N. 1978. Regulation of blood pH in teleost fish. *Resp. Physiol.* **33**, 129–44.

Campbell, J.W. & Speeg, K.V. 1968. Arginine biosynthesis and metabolism in terrestrial snails. *Comp. Biochem. Physiol.* **25**, 3–32.

Carley, W.A. 1978. Water economy of the earthworm *Lumbricus terrestris* L.: coping with the terrestrial environment. *J. exp. Zool.* **205**, 71–8.

Carroll, R.L. 1970. Quantitative aspects of the amphibian-reptilian transition. *Forma et Functio* **3**, 165–78.

    1980. The hyomandibular as a supporting element in the skull of primitive tetrapods. In *The terrestrial environment and the origin of land vertebrates*, ed. A.L. Panchen, pp. 293–317. London: Academic Press.

    1988. *Vertebrate paleontology and evolution*. New York: W.H. Freeman.

Carter, G.S. & Beadle, L.C. 1931a. The fauna of the swamps of the Paraguayan Chaco in relation to its environment. I. Physico-chemical nature of the environment. *J. Linn. Soc. (Zool.)* **37**, 205–58.

    1931b. The fauna of the swamps of the Paraguayan Chaco in relation to its environment. II. Respiratory adaptations in the fishes. *J. Linn. Soc. (Zool.)* **37**, 327–68.

Chaloner, W.G. 1970. The rise of the first land plants. *Biol. Rev.* **45**, 353–77.

Chaloner, W.G. & Lacey, W.S. 1973. The distribution of late Palaeozoic floras. In *Organisms and continents through time*, ed. N.F. Hughes. *Spec. Pap. Palaeont.* **12**, 271–89.

Chapman, V.J. 1960. The plant ecology of Scolt Head Island. In *Scolt Head Island*, ed. J.A. Steers, pp. 85–163. Cambridge: W. Heffer & Sons, revised edition.

    1977a. Introduction. In *Ecosystems of the world, vol. 1, Wet coastal ecosystems*, ed. V.J. Chapman, pp. 1–29. New York: Elsevier.

    1977b. Wet coastal formations of Indo-Malesia and Papua-New Guinea. In *Ecosystems of the world, vol. 1, Wet coastal ecosystems*, ed. V.J. Chapman, pp. 261–70. New York: Elsevier.

Chesters, K.I.M., Gnauck, F.R. & Hughes, N.F. 1967. Angiospermae. In *The fossil record*, ed. W.B. Harland *et al.*, pp. 269–88. London: Geological Society of London.

Christy, J.H. & Salmon, M. 1984. Ecology and evolution of mating systems of fiddler crabs (genus *Uca*). *Biol. Rev.* **59**, 483–509.

Coenen-Stass, D. 1981. Some aspects of the water balance of two desert woodlice, *Hemilepistus aphganicus* and *Hemilepistus reaumuri* (Crustacea, Isopoda, Oniscoidea). *Comp. Biochem. Physiol.* **70A**, 405–19.

Conway Morris, S.C. 1986. The community structure of the Middle Cambrian phyllopod bed (Burgess Shale). *Palaeontology*, **29**, 423–67.

Cooke, A.H. 1895. Molluscs. In *The Cambridge natural history*, eds. S.F. Harmer & A.E. Shipley, vol. III, pp. 1–459. London: Macmillan & Co.

Daiber, F.C. 1982. *Animals of the tidal marsh*. New York: Van Nostrand Reinhold Co.

Dainton, B.H. 1954. The activity of slugs. I. The induction of activity by changing temperatures. *J. exp. Biol.* **31**, 165–87.

Daniel, M.J. & Boyden, C.R. 1975. Diurnal variations in physico-chemical conditions within rockpools. *Fld. Stud.* **4**, 161–76.

Darwin, C.R. 1881. *The formation of vegetable mould through the action of worms, with observations on their habits*. London: John Murray.

Davies, L. & Richardson, J. 1970. Distribution in Britain and habitat requirements of *Petrobius maritimus* (Leach) and *P. brevistylis* Carpenter (Thysanura). *Entomologist* **103**, 97–114.

Decoursey, P.J. 1983. Biological timing. In *The Biology of Crustacea*, eds. F.J. Vernberg & W.B. Vernberg, vol. 7, pp. 107–62. London: Academic Press.

Dehadrai, P.V. & Tripathi, S.D. 1976. Environment and ecology of freshwater air-breathing teleosts. In *Respiration of amphibious vertebrates*, ed. G.M. Hughes, pp. 39–72. London: Academic Press.

Dejours, P. 1988. *Respiration in water and air. Adaptations – regulation – evolution*. Amsterdam: Elsevier.

Delany, M.J. 1953. Studies on the microclimate of *Calluna* heathland. *J. Anim. Ecol.* **22**, 227–39.

1959. The life histories and ecology of two species of *Petrobius* Leach, *P. brevistylis* and *P. maritimus*. *Trans. R. Soc. Edinb.* **63**, 501–33.

Delevoryas, T. 1977. *Plant diversification* (2nd edition). New York: Holt, Rhinehart & Winston.

Delhaye, W. & Bouillon, J. 1972. L'évolution et l'adaptation de l'organe excréteur chez les mollusques gastéropodes pulmonés. II. Histophysiologie comparée du rein chez les stylommatophores. *Bull. bio. Fr. Belg.* **106**, 123–42.

Dietz, R.S. & Holden, J.C. 1976. The breakup of Pangaea. In *Continents adrift and continents aground*, ed. J.T. Wilson, pp. 126–37. San Francisco: W.H. Freeman.

Dietz, T.H. 1974. Active chloride transport across the skin of the earthworm, *Lumbricus terrestris* L. *Comp. Biochem. Physiol.* **49A**, 251–8.

Digby, P.S.B. 1955. Factors affecting the temperature excess of insects in sunshine. *J. exp. Biol.* **32**, 279–98.

Edney, E.B. 1953. The temperature of woodlice in the sun. *J. exp. Biol.* **30**, 331–49.

1957. *The water relations of terrestrial arthropods*. Cambridge University Press.

1968. Transition from water to land in isopod crustaceans. *Am. Zool.* **8**, 309–26.

1971. The body temperature of tenebrionid beetles in the Namib desert of southern Africa. *J. exp. Biol.* **55**, 253–72.

1977. *Water balance in land arthropods*. New York: Springer-Verlag.

Edney, E.B. & Spencer, J. 1955. Cutaneous respiration in woodlice. *J. exp. Biol.* **32**, 256–69.

Edwards, C.A. & Lofty, J.R. 1972. *Biology of earthworms*. London: Chapman & Hall.

Edwards, D. 1980. Early land floras. In *The terrestrial environment and the origin of land vertebrates*, ed. A.L. Panchen, pp. 55–85. Systematics Association Special Volume

15. London: Academic Press.

Edwards, D., & Fanning, U. 1985. Evolution and environment in the late Silurian-early Devonian: the rise of the pteridophytes. *Phil. Trans. R. Soc. Lond.* **B309**, 147–65.

Edwards, J.L. 1976. The evolution of terrestrial locomotion. In *Major patterns in vertebrate evolution*, eds. M.K. Hecht, P.C. Goody & B.M. Hecht, pp. 553–77. London: Plenum Press.

Edwards, P.J. & Wratten, S.D. 1980. *Ecology of insect-plant interactions.* London: Arnold.

Erséus, C. 1980. Specific and generic criteria in marine Oligochaeta, with special emphasis on Tubificidae. In *Aquatic oligochaete biology*, eds. R.O. Brinkhurst & D.G. Cook, pp. 9–24. London: Plenum Press.

    1984. Aspects of the phylogeny of the marine Tubificidae. *Hydrobiologia* **115**, 37–44.

Evans, D.H. 1979. Fish. In *Comparative physiology of osmoregulation in animals Vol. 1*, ed. G.M.O. Maloiy, pp. 305–90. London: Academic Press.

Farber, J. & Rahn, H. 1970. Gas exchange between air and water and the ventilation pattern of the electric eel. *Resp. Physiol.* **9**, 151–61.

Fielding, P.J. & Nicolson, S.W. 1980. Regulation of haemolymph osmoregularity and ions in the green protea beetle, *Trichostetha fascicularis*, during dehydration and rehydration. *Comp. Biochem. Physiol.* **67A**, 691–4.

Fitzpatrick, E.A. 1980. *Soils. Their formation, classification and distribution.* London: Longman.

Flemister, L.J. 1958. Salt and water anatomy, constancy and regulation in related crabs from marine and terrestrial habitats. *Biol. Bull. mar. biol. Lab., Woods Hole* **115**, 180–200.

Foissner, W. 1987. Soil Protozoa: fundamental problems, adaptations in ciliates and testaceans, bioindicators, and a guide to the literature. *Progress in Protistology* **2**, 69–212. Bristol: Biopress Ltd.

Fretter, V. & Graham, A. 1962. *British Prosobranch Molluscs.* London: Ray Society.

    1980. The prosobranch molluscs of Britain and Denmark. Part 5 – Marine Littorinacea. *J. Moll. Stud.*, supplement **7**, 243–84.

Fretter, V. & Manly, R. 1977. Algal associations of *Tricolia pullus*, *Lacuna vincta* and *Cerithiopsis tubercularis* (Gastropoda) with special reference to the settlement of their larvae. *J. mar. biol. Ass. U.K.* **57**, 999–1017.

Friend, J.A. & Richardson, A.M.M. 1986. Biology of terrestrial amphipods. *Ann. Rev. Entomol.* **31**, 25–48.

Gabe, M. 1957. Données histologiques sur les organes segmentaires des Peripatopsidae (Onychophores). *Arch. Anat. microsc. Morph. exp.* **46**, 283–306.

Gans, C. 1970. Respiration in the early tetrapods: the frog is a red herring. *Evolution* **24**, 723–34.

Gardiner, B.G. 1980. Tetrapod ancestry: a reappraisal. In *The terrestrial environment and the origin of land vertebrates*, ed. A.L. Panchen, pp. 177–85. London: Academic Press.

Garrity, S.D. 1984. Some adaptations of gastropods to physical stress on a tropical rocky shore. *Ecology* **65**, 559–74.

Geiger, R. 1965. *Das Klima der bodennahen Luftschicht.* Brunswick, Germany: F. Viewig & Sohn. Translated as *The climate near the ground*, Cambridge, Mass: Harvard University Press.

Ghilarov, M.S. 1959. Adaptations of insects to soil dwelling. In *Proceedings of the XVth international congress of zoology, London*, pp. 354–7.

Ghiretti, F. & Ghiretti-Magaldi, A. 1975. Respiration. In *Pulmonates*, eds. K.M. Wilbur & C.M. Yonge, vol. 2, pp. 233–48. London: Academic Press.

Gibson, R., Moore, J. & Crandall, F.B. 1982. A new semi-terrestrial nemertean from

California. *J. Zool. (Lond.)* **196**, 463–74.

Giere, O. & Pfannkuche, O. 1982. Biology and ecology of marine Oligochaeta, a review. *Oceanogr. mar. Biol. ann. Rev.* **20**, 173–308.

Gifford, C.A. 1962. Some observations on the general biology of the land crab, *Cardisoma guanhumi* (Latreille), in south Florida. *Biol. Bull. mar. biol. Lab., Woods Hole* **132**, 207–23.

Gilbert, L.E. 1982. The coevolution of a butterfly and a vine. *Scient. Am.* **247**, 102–7.

Glynne-Williams, J. & Hobart, J. 1952. Studies on the crevice fauna of a selected shore in Anglesey. *Proc. zool. Soc. Lond.* **122**, 794–824.

Gordon, M.S., Ng, W.W. & Yip, A.Y. 1978. Aspects of terrestrial life in amphibious fishes. III. The Chinese mudskipper, *Periophthalmus cantonensis. J. exp. Biol.* **72**, 57–75.

Gordon, M.S., Schmidt-Nielsen, K. & Kelly, H.M. 1961. Osmotic regulation in the crab-eating frog (*Rana cancrivora*). *J. exp. Biol.* **38**, 659–78.

Gordon, M.S. & Tucker, V.A. 1965. Osmotic regulation in the tadpoles of the crab-eating frog (*Rana cancrivora*). *J. exp. Biol.* **42**, 437–45.

Graham, J.B. & Baird, T.A. 1982. The transition to air breathing in fishes. I. Environmental effects on the facultative air breathing of *Ancistrus chagresi* and *Hypostomus plecostomus* (Loricariidae). *J. exp. Biol.* **96**, 53–67.

Gray, A.J. 1985. Adaptation in perennial coast plants – with particular reference to heritable variation in *Puccinellia maritima* and *Ammophila arenaria. Vegetatio* **61**, 179–88.

Gray, J. 1968. *Animal Locomotion*. London: Weidenfeld & Nicolson.

1985a. The microfossil record of early land plants: advances in understanding of early terrestrialization, 1970–1984. *Phil. Trans. R. Soc. Lond.* **B309**, 167–95.

1985b. Ordovician-Silurian land plants: the interdependence of ecology and evolution. *Spec. Pap. Palaeontol.* **32**, 281–95.

1988. Evolution of the freshwater ecosystem: the fossil record. *Palaeogeography, Palaeoclimatology, Palaeoecology* **62**, 1–214.

Gray, J.S. 1981. *The ecology of marine sediments*. Cambridge University Press.

Greenaway, P. & MacMillen, R.E. 1978. Salt and water balance in the terrestrial phase of the inland crab *Holthuisana (Austrothelphusa) transversa* Martens (Parathelphusoidea: Sundathelphusidae). *Physiol. Zool.* **51**, 217–29.

Greenaway, P., Morris, S. & McMahon, B.R. 1988. Adaptations to a terrestrial existence by the robber crab *Birgus latro*. II. *In vivo* respiratory gas exchange and transport. *J. exp. Biol.* **140**, 493–509.

Griffith, R.W. 1987. Freshwater or marine origin of the vertebrates? *Comp. Biochem. Physiol.* **87A**, 523–31.

Gross, W.J. 1957. A behavioural mechanism for osmotic regulation in a semi-terrestrial crab. *Biol. Bull. mar. biol. Lab., Woods Hole* **113**, 268–74.

1964. Trends in water and salt regulation among aquatic and amphibious crabs. *Biol. Bull. mar. biol. Lab., Woods Hole* **127**, 447–66.

Gross, W.J. & Holland, P.V. 1960. Water and ionic regulation in a terrestrial hermit crab. *Physiol. Zoöl.* **33**, 21–8.

Halstead, L.B. 1973. The heterostracan fishes. *Biol. Rev.* **48**, 279–332.

Hamilton, P.V. 1978. Adaptive visually-mediated movements of *Littorina irrorata* (Mollusca: Gastropoda) when displaced from their natural habitat. *Mar. Behav. Physiol.* **5**, 255–71.

Hamilton, P.V., Ardizzoni, S.C. & Penn, J.S. 1983. Eye structure and optics in the intertidal snail, *Littorina irrorata. J. comp. Physiol.* **152A**, 435–45.

Hamilton, W.J. & Seely, M.K. 1976. Fog basking by the Namib desert beetle, *Onymacris unguicularis. Nature, Lond.* **262**, 284–5.

Hardy, R.N. 1972. *Temperature and animal life*. London: Arnold.

Harrison, S.J. & Phizacklea, A.P. 1987. Temperature fluctuation in muddy intertidal sediments, Forth Estuary, Scotland. *Est. coast. Shelf Sci.* **24**, 279–88.

Hartnoll, R.G. 1963. The freshwater grapsid crabs of Jamaica. *Proc. Linn. Soc., Lond.* **175**, 145–69.

  1973. Factors affecting the distribution and behaviour of the crab *Dotilla fenestrata* on east African shores. *Est. coast. mar. Sci.* **1**, 137–52.

Hawkins, A.J.S. & Jones, M.B. 1982. Gill area and ventilation in two mud crabs, *Helice crassa* Dana (Grapsidae) and *Macrophthalmus hirtipes* (Jacquinot) (Ocypodidae), in relation to habitat. *J. exp. mar. Biol. Ecol.* **60**. 103–18.

Healy, B. & Bolger, T. 1984. The occurrence of species of semi-aquatic Enchytraeidae (Oligochaeta) in Ireland. *Hydrobiologia* **115**, 159–70.

Heinrich, B. 1975. Thermoregulation and flight energetics of desert insects. In *Environmental physiology of desert organisms*, ed. N.F. Hadley, pp. 90–105. Stroudsberg, Pennsylvania: Dowden, Hutchinson & Ross Inc.

Henwood, K. 1975. A field-tested thermoregulation model for two diurnal Namib Desert tenebrionid beetles. *Ecology* **56**, 1329–42.

Herreid, C.F. 1969. Integument permeability of crabs and adaptation to land. *Comp. Biochem. Physiol.* **29**, 423–9.

Herreid, C.F. & Gifford, C.A. 1963. The burrow habitat of the land crab *Cardisoma guanhumi* Latreille. *Ecology* **44**, 773–5.

Herrnkind, W.F. 1983. Movement patterns and orientation. In *The biology of Crustacea*, eds. F.J. Vernberg & W.B. Vernberg, vol. 7, pp. 41–105. London: Academic Press.

Hiatt, R.W. 1948. The biology of the lined shore crab, *Pachygrapsus crassipes* Randall. *Pacific Sci.* **2**, 135–213.

Hinton, H.E. 1968. Reversible suspension of metabolism and the origin of life. *Proc. roy. Soc. Lond.* (B) **171**, 43–57.

  1977. Enabling mechanisms. *Proc. Int. Congr. Ent.* **15**, 71–83.

Hinton, H.E. & Blum, M.S. 1965. Suspended animation and the origin of life. *New Scient.* **28** (467), 270–1.

Hobbs, H.H. & Rewolinski, S.A. 1985. Notes on the burrowing crayfish *Procambarus (Girardiella) gracilis* (Bundy) (Decapoda, Cambaridae) from southeastern Wisconsin, U.S.A. *Crustaceana* **48**, 26–33.

Hoese, B. 1981. Morphologie und Funktion des Wasserleitungssystems der terrestrischen Isopoden (Crustacea, Isopoda, Oniscoidea). *Zoomorphol.* **98**, 135–67.

  1983. Struktur und Entwicklung der Lungen der Tylidae (Crustacea, Isopoda, Oniscoidea). *Zool. Jb. Anat.* **109**, 487–501.

Holanov, S.H. & Hendrickson, J.R. 1980. The relationship of sand moisture to burrowing depth of the sand-beach isopod *Tylos punctatus* Holmes and Gray. *J. exp. mar. Biol. Ecol.* **46**, 81–8.

Holdich, D.M., Lincoln, R.J. & Ellis, J.P. 1984. The biology of terrestrial isopods: terminology and classification. In *The biology of terrestrial isopods*, eds. S.L. Sutton & D.M. Holdich, pp. 1–6. Oxford: Clarendon Press.

Horch, K. 1971. An organ for hearing and vibration sense in the ghost crab *Ocypode*. *Z. vergl. Physiol.* **73**, 1–21.

Horowitz, M. 1970. The water balance of the terrestrial isopod *Porcellio scaber*. *Ent. exp. appl.* **13**, 173–8.

Horwitz, P.H.J. & Richardson, A.M.M. 1986. An ecological classification of the burrows of Australian freshwater crayfish. *Aust. J. mar. freshwat. Res.* **37**, 237–42.

Horwitz, P.H.J., Richardson, A.M.M. & Boulton, A. 1985. The burrow habitat of two sympatric species of land crayfish, *Engaeus urostrictus* and *E. tuberculatus* (Decapoda: Parastacidae). *Victorian Naturalist* **102**, 188–97.

Houlihan, D.F. 1976. Water transport by the eversible abdominal vesicles of *Petrobius brevistylis*. *J. Insect Physiol.* **22**, 1683–95.

1979. Respiration in air and water of three mangrove snails. *J. exp. mar. Biol. Ecol.* **41**, 143–61.

Howes, N.H. & Wells, G.P. 1934. The water relations of snails and slugs. I. Weight rhythms in *Helix pomatia* L. *J. exp. Biol.* **11**, 327–43.

Huebner, E. & Chee, G. 1978. Histological and ultrastructural specialization of the digestive tract of the intestinal air breather *Hoplosternum thoracatum* (Teleost). *J. Morph.* **157**, 301–28.

Hughes, D.A. 1966. Behavioural and ecological investigations of the crab *Ocypode ceratophthalma* (Crustacea: Ocypodidae). *J. Zool., Lond.* **150**, 129–43.

Hughes, G.M. 1967. Evolution between air and water. In *Development of the lung*, eds. A.V.S. de Reuck & R. Porter, pp. 64–84. London: J. & A. Churchill Ltd.

Hughes, N.F. 1976. *Palaeobiology of angiosperm origins*. Cambridge University Press.

Hughes, N.F. & Smart, J. 1967. Plant-insect relationships in Palaeozoic and later time. In *The fossil record*, eds. W.B. Harland, *et al.*, pp. 107–17. London: Geological Society.

Hutchinson, G.E. 1975. *A treatise on limnology, vol. I, Geography, physics and chemistry*. London: John Wiley & Sons.

Hutchings, P.A. & Recher, H.F. 1982. The fauna of Australian mangroves. *Proc. Linn. Soc. N.S.W.* **106**, 83–121.

Huxley, C.R. 1986. Evolution of benevolent ant-plant relationships. In *Insects and the plant surface*, eds. B. Juniper & T.R.E. Southwood, pp. 257–82. London: Arnold.

Hynes, H.B.N. 1970. *The ecology of running waters*. Liverpool University Press.

Inger, R.F. 1957. Ecological aspects of the origin of the tetrapods. *Evolution* **11**. 373–6.

Innes, A.J., Forster, M.E., Jones, M.B., Marsden, I.D. & Taylor, H.H. 1986. Bimodal respiration, water balance and acid-base regulation in a high-shore crab, *Cyclograpsus lavauxi* H. Milne Edwards. *J. exp. mar. Biol. Ecol.* **100**, 127–45.

Innes, A.J. & Taylor, E.W. 1986. The evolution of air-breathing in crustaceans: a functional analysis of branchial, cutaneous and pulmonary gas exchange. *Comp. Biochem. Physiol.* **85A**, 621–37.

Innes, A.J., Taylor, E.W. & Haj, A.J. El. 1987. Air-breathing in the Trinidad mountain crab: a quantum leap in the evolution of the invertebrate lung? *Comp. Biochem. Physiol.* **87A**, 1–8.

Janzen, D.H. 1980. When is it coevolution? *Evolution* **34**, 611–12.

Jarvik, E. 1980. *Basic structure and evolution of vertebrates*. 2 vols. London: Academic Press.

1981. Review of 'Lungfishes, tetrapods, paleontology, and plesiomorphy'. *Syst. Zool.* **30**, 378–84.

Jefferies, R.P.S. 1986. *The ancestry of the vertebrates*. London: British Museum (Natural History).

Jennings, J.N. 1965. Further discussion of factors affecting coastal dune formation in the tropics. *Aust. J. Sci.* **28**, 166–7.

Johansen, K. & Hanson, D. 1968. Functional anatomy of the hearts of lungfishes and amphibians. *Am. Zool.* **8**, 191–210.

Johansen, K., Lenfant, C., & Hanson, D. 1968. Cardiovascular dynamics in the lungfish. *Z. vergl. Physiol.* **59**, 157–86.

Johansen, K., Lomholt, J.P. & Maloiy, G.M.O. 1976. Importance of air and water breathing in relation to size of the African lungfish *Protopterus amphibius* Peters. *J. exp. Biol.* **65**, 395–9.

Johnels, A.G. & Svensson, G.S.O. 1954. On the biology of *Protopterus annectens* (Owen). *Ark Zool.* **7**, 131–64.

Jollie, M. 1982. What are the 'Calcichordata'? and the larger question of the origin of chordates. *Zool. J. Linn. Soc.* **75**, 167–88.

Jones, H.D. 1978. Observations on the locomotion of two British terrestrial planarians (Platyhelminthes, Tricladida). *J. Zool., Lond.* **186**, 407–16.

Jones, J.D. 1961. Aspects of respiration in *Planorbis corneus* L. and *Lymnaea stagnalis* L. (Gastropoda: Pulmonata). *Comp. Biochem. Physiol.* **4**, 1–29.

Jones, M.B. & Simons, M.J. 1982. Habitat preferences of two estuarine burrowing crabs *Helice crassa* Dana (Grapsidae) and *Macrophthalmus hirtipes* (Jacquinot) (Ocypodidae). *J. exp. mar. Biol. Ecol.* **56**, 49–62.

Jurgens, J.D. 1971. The morphology of the nasal region of Amphibia and its bearing on the phylogeny of the group. *Ann. Univ. Stellenbosch* **46A**, 1–146.

Kaufman, S.E., Kaufman, W.R. & Phillips, J.E. 1982. Mechanism and characteristics of coxal fluid excretion in the argasid tick *Ornithodorus moubata. J. exp. Biol.* **98**, 343–52.

Kemp, T.S. 1980. Origin of the mammal-like reptiles. *Nature, Lond.* **283**, 378–80.

Kitching, R.L. 1986. Prey-predator interactions. In *Community ecology: pattern and process*, eds. J. Kikkawa & D.J. Anderson, pp. 214–39. Oxford: Blackwell.

Kneib, R.T. 1982. Habitat preference, predation, and the intertidal distribution of gammaridean amphipods in a North Carolina salt marsh. *J. exp. mar. Biol. Ecol.* **59**, 219–30.

Kummel, G. 1975. The physiology of protonephridia. *Fortschr. Zool.* **23**, 18–32.

Lange, R. 1972. Some recent work on osmotic, ionic and volume regulation in marine animals. *Oceanogr. mar. Biol. Ann. Rev.* **10**, 97–136.

Lavallard, R., Campiglia, S., Parisi-Alvares, E. & Valle, C.M.C. 1975. Contribution a la biologie de *Peripatus acacioi* Marcus et Marcus. III. Etude descriptive de l'habitat. *Vie et Milieu* **25**, 87–118.

Lawton, J.H. 1986. Surface availability and insect community structure: the effects of architecture and fractal dimension of plants. In *Insects and the plant surface* eds. B. Juniper & T.R.E. Southwood, pp. 317–31. London: Arnold.

Lazo-Wasem, E.A. 1984. Physiological and behavioral ecology of the terrestrial amphipod *Arcitalitrus sylvaticus* (Haswell, 1880). *J. Crustacean Biol.* **4**, 343–55.

Levinton, J.S. 1982. *Marine ecology.* Englewood Cliffs, New Jersey: Prentice-Hall Inc.

Lewis, J.R. 1964. *The ecology of rocky shores.* London: English Universities Press.

Lillywhite, H.B. & Maderson, P.F.A. 1988. The structure and permeability of integument. *Am. Zool.* **28**, 945–62.

Lissmann, H.W. 1945. The mechanism of locomotion in gastropod molluscs. I. Kinematics. *J. exp. Biol.* **21**, 58–69.

Little, C. 1965. Osmotic and ionic regulation in the prosobranch gastropod mollusc, *Viviparus viviparus* Linn. *J. exp. Biol.* **43**, 39–54.

1968. Aestivation and ionic regulation in two species of *Pomacea* (Gastropoda, Prosobranchia). *J. exp. Biol.* **48**, 569–85.

1972. The evolution of kidney function in the Neritacea (Gastropoda, Prosobranchia). *J. exp. Biol.* **56**, 249–61.

1981. Osmoregulation and excretion in prosobranch gastropods. Part I. Physiology and biochemistry. *J. moll. Stud.* **47**, 221–47.

1983. *The colonisation of land. Origins and adaptations of terrestrial animals.* Cambridge University Press.

1984. Ecophysiology of *Nais elinguis* (Oligochaeta) in a brackish-water lagoon. *Est. coast. Shelf Sci.* **18**, 231–44.

1985. Renal adaptations of prosobranchs to the freshwater environment. *Am. malacol. Bull.* **3**, 223–31.

1989a. Factors governing patterns of foraging activity in littoral marine herbivorous molluscs. *J. moll. Stud.* **55**, 273–84.

1989b. Comparative physiology as a tool for investigating the evolutionary routes of animals on to land. *Trans. roy. Soc. Edinb. Earth Sciences,* **80**, 201–8.

Little, C. & Andrews, E.B. 1977. Some aspects of excretion and osmoregulation in assimineid snails. *J. moll. Stud.* **43**, 265–85.

Little, C., Pilkington, M.C. & Pilkington, J.B. 1984. Development of salinity tolerance in the marine pulmonate *Amphibola crenata* (Gmelin). *J. exp. mar. Biol. Ecol.* **74**, 169–77.

Little, C. & Stirling, P. 1985. Activation of a mangrove snail, *Littorina scabra scabra* (L.) (Gastropoda: Prosobranchia). *Aust. J. mar. fw. Res.* **35**, 607–10.

Little, C., Stirling, P., Pilkington, M. & Pilkington, J. 1985. Larval development and metamorphosis in the marine pulmonate *Amphibola crenata* (Mollusca: Pulmonata). *J. Zool., Lond.* **205**, 489–510.

Little, C., Williams, G.A., Morritt, D. & Seaward, D.R. 1989. Distribution of intertidal molluscs in lagoonal shingle (The Fleet, Dorset). J. Conch., in press.

Livermore, R.A., Smith, A.G. & Briden, J.C. 1985. Palaeomagnetic constraints on the distribution of continents in the late Silurian and early Devonian. *Phil. Trans. R. Soc.* **B309**, 29–56.

Lombard, R.E. & Bolt, J.R. 1979. Evolution of the tetrapod ear: an analysis and reinterpretation. *Biol. J. Linn. Soc.* **11**, 19–76.

Long, S.P. & Mason, C.F. 1983. *Saltmarsh ecology*. Glasgow: Blackie.

Luckett, W.P. 1976. Ontogeny of amniote fetal membranes and their application to phylogeny. In *Major patterns in vertebrate evolution* eds. M.K. Hecht, P.C. Goody & B.M. Hecht, pp. 439–516. London: Plenum Press.

Lutz, P.L. 1969. Salt and water balance in the West African freshwater/land crab *Sudanonautes africanus africanus* and the effects of desiccation. *Comp. Biochem. Physiol.* **30**, 469–80.

Macallum, A.B. 1926. The paleochemistry of the body fluids and tissues. *Physiol. Rev.* **6**, 316–57.

McClanahan, L. 1967. Adaptation of the spadefoot toad, *Scaphiopus couchii*, to desert environments. *Comp. Biochem. Physiol.* **20**, 73–99.

McClary, A. 1964. Surface inspiration and ciliary feeding in *Pomacea paludosa* (Prosobranchia: Mesogastropoda: Ampullariidae). *Malacologia* **2**, 87–104.

McGhee G.R. 1989. Frasnian-Famennian extinction event. In *Palaeobiology – a synthesis*, eds. D.E.G. Briggs & P.R. Crowther. Oxford: Blackwell Scientific Publications.

Mackenzie, F.T. 1975. Sedimentary cycling and the evolution of sea water. In *Chemical oceanography* (second edition), eds. J.P. Riley & G. Skirrow, vol. I, pp. 309–64. London: Academic Press.

McLachlan, A. 1978. A quantitative analysis of the meiofauna and the chemistry of the redox potential discontinuity zone in a sheltered sandy beach. *Est. coast. mar. Sci.* **7**, 275–90.

   1983. Sandy beach ecology – a review. In *Sandy beaches as ecosystems,* eds. A. McLachlan & T. Erasmus, pp. 321–80. The Hague: Dr W. Junk.

   1985. The biomass of macro- and interstitial fauna on clean and wrack-covered beaches in western Australia. *Est. coast. Shelf Sci.* **21**, 587–99.

McLusky, D.S. 1981. *The estuarine ecosystem*. Glasgow: Blackie.

McMahon, B.R. & Burggren, W.W. 1987. Respiratory physiology of intestinal air breathing in the teleost fish *Misgurnus anguillicaudatus*. *J. exp. Biol.* **133**, 371–93.

McMahon, B.R. & Wilkens, J.L. 1983. Ventilation, perfusion and oxygen uptake. In *The physiology of Crustacea*, ed. L.H. Mantel, vol. 5, pp. 289–372. London: Academic Press.

McMahon, R.F. 1988. Respiratory response to periodic emergence in intertidal molluscs. *Am. Zool.* **28**, 97–114.

McMahon, R.F. & Britton, J.C. 1985. The relationship between vertical distribution, thermal tolerance, evaporative water loss rate, and behaviour on emergence in six species of mangrove gastropods from Hong Kong. In *Proceedings of the second international workshop on the malacofauna of Hong Kong and southern China, Hong Kong, 1983*, ed. B. Morton & D. Dudgeon, pp. 563–82. Hong Kong: Hong Kong University Press.

MacMillen, R.E. & Greenaway, P. 1978. Adjustments of energy and water metabolism to drought in an Australian arid-zone crab. *Physiol. Zool.* **51**, 231–40.

Macnae, W. 1968. A general account of the fauna and flora of mangrove swamps and forests in the Indo-West-Pacific region. *Adv. mar. Biol.* **6**, 73–270.

Macnae, W. & Kalk, M. 1962. The ecology of the mangrove swamps at Inhaca island, Mocambique. *J. Ecol.* **50**, 19–34.

Machin, J. 1966. The evaporation of water from *Helix aspersa*. IV. Loss from the mantle of the inactive snail. *J. exp. Biol.* **45**, 269–78.

1975. Water relationships. In *Pulmonates*, eds. V. Fretter & J. Peake, vol. 1, pp. 105–163. London: Academic Press.

1977. Role of integument in molluscs. In *Transport of ions and water in animals*, eds. B.L. Gupta, R.B. Moreton, J.L. Oschman & B.J. Wall, pp. 735–62. London: Academic Press.

Maddrell. S.H.P. 1981. The functional design of the insect excretory system. *J. exp. Biol.* **90**, 1–15.

Maitland, D.P. 1986. Crabs that breathe air with their legs – *Scopimera* and *Dotilla*. *Nature, Lond.* **319**, 493–5.

Mangum, C.P. 1983. Oxygen transport in the blood. In *The physiology of Crustacea*, ed. L.H. Mantel, vol. 5, pp. 373–429. London: Academic Press.

Mangum, C.P., Lykkeboe, G. & Johansen, K. 1975. Oxygen uptake and the role of hemoglobin in the east African swampworm *Alma emini*. *Comp. Biochem. Physiol.* **52A**, 477–82.

Mann, K.H. 1962. *Leeches (Hirudinea). Their structure, physiology, ecology and embryology*. Oxford: Pergamon Press.

Mantel, L.H. & Farmer, L.L. 1983. Osmotic and ionic regulation. In *The biology of Crustacea Vol. 5*, ed. L.H. Mantel, pp. 53–161. London: Academic Press.

Manton, S.M. 1952a. The evolution of arthropodan locomotory mechanisms – Part 2. General introduction to the locomotory mechanisms of the Arthropoda. *J. Linn. Soc.* **42**, 93–117.

1952b. The evolution of arthropodan locomotory mechanisms – Part 3. The locomotion of the Chilopoda and Pauropoda. *J. Linn. Soc* **42**, 118–67.

1977. *The Arthropoda. Habits, functional morphology and evolution*. Oxford University Press.

1979. Functional morphology and the evolution of the hexapod classes. In *Arthropod phylogeny*, ed. A.P. Gupta, pp. 387–467. New York: Van Nostrand Reinhold Co.

Marsden, I.D. 1976. Effect of temperature on the microdistribution of the isopod *Sphaeroma rugicauda* from a saltmarsh habitat. *Mar. Biol.* **38**, 117–28.

1980. Effects of constant and cyclic temperatures on the salinity tolerance of the estuarine sandhopper *Orchestia chiliensis*. *Mar. Biol.* **59**, 211–18.

Marsh, B.A. & Branch, G.M. 1979. Circadian and circatidal rhythms of oxygen consumption in the sandy-beach isopod *Tylos granulatus* Krauss. *J. exp. mar. Biol. Ecol.* **37**, 77–89.

Marshall, N.B. 1965. *The life of fishes*. London: Weidenfeld and Nicolson.

Matthews, E.G. 1976. *Insect ecology*. St Lucia, Queensland: University of Queensland Press.

May, R.M. 1978. The dynamics and diversity of insect faunas. In *Diversity of insect faunas*, eds. L.A. Mound & N. Waloff, pp. 188–204. *Symp. R. Ent. Soc. Lond.* **9**. Oxford: Blackwell.

Minnich, J.E. 1979. Reptiles. In *Comparative physiology of osmoregulation in animals*, ed. G.M.O. Maloiy, vol. 1, pp. 391–641. London: Academic Press.

Moore, J. 1973. Land nemertines of New Zealand. *Zool. J. Linn. Soc.* **52**, 293–313.

Moore, J. & Gibson, R. 1973. A new genus of freshwater hoplonemerteans from New Zealand. *Zool. J. Linn. Soc.* **52**, 141–57.

1981. The *Geonemertes* problem (Nemertea). *J. Zool., Lond.* **194**, 175–201.

1985. The evolution and comparative physiology of terrestrial and freshwater nemerteans. *Biol. Rev.* **60**, 257–312.

Moore, P.G. & Francis, C.H. 1985. On the water relations and osmoregulation of the beach-hopper *Orchestia gammarellus* (Pallas) (Crustacea: Amphipoda). *J. exp. mar. Biol. Ecol.* **94**, 131–50.

Moore, P.G. & Powell, H.T. 1985. J.R. Lewis and the ecology of rocky shores. In *The ecology of rocky coasts*, eds. P.G. Moore & R. Seed, pp. 1–6. London: Hodder & Stoughton.

Moore, P.G. & Taylor, A.C. 1984. Gill area relationships in an ecological series of gammaridean amphipods (Crustacea). *J. exp. mar. Biol. Ecol.* **74**, 179–86.

Morris, M.G. 1968. Differences between the invertebrate faunas of grazed and ungrazed chalk grassland. II. The faunas of sample turves. *J. appl. Ecol.* **5**, 601–11.

Morritt, D. 1987. Evaporative water loss under desiccation stress in semiterrestrial and terrestrial amphipods (Crustacea: Amphipoda: Talitridae). *J. exp. mar. Biol. Ecol.* **111**, 145–57.

1988. Osmoregulation in littoral and terrestrial talitroidean amphipods (Crustacea) from Britain. *J. exp. mar. Biol. Ecol.* **123**, 77–94.

Morton, J.E. 1954. The crevice fauna of the upper intertidal zone at Wembury. *J. mar. biol. Ass. U.K.* **33**, 187–224.

1955. The evolution of the Ellobiidae with a discussion on the origin of the Pulmonata. *Proc. zool. Soc. Lond.* **125**, 127–68.

Morton, J.E. & Miller, M. 1973. *The New Zealand Sea Shore*, 2nd edn. London: Collins.

Munshi, J.S.D. 1976. Gross and fine structure of the respiratory organs of air-breathing fishes. In *Respiration of amphibious vertebrates*, ed. G.M. Hughes, pp. 73–104. London: Academic Press.

Mullins, D.E. & Cochran, D.G. 1973. Nitrogenous excretory materials from the American cockroach. *J. Insect Physiol.* **19**, 1007–18.

Newell, R.C. 1970. *Biology of intertidal animals*. London: Paul Elek Ltd.

Nielsen, C. 1976. Notes on *Littorina* and *Murex* from the mangrove at Ao Nam-bor, Phuket, Thailand. *Res. Bull. Phuket Mar. Biol Center* **11**, 1–4.

Nørgaard, E. 1956. Environment and behaviour of *Theridion saxatile*. *Oikos* **7**, 159–92.

Norton, T.A. 1985. The zonation of seaweeds on rocky shores. In *The ecology of rocky coasts*, eds. P.G. Moore & R. Seed, pp. 7–21. London: Hodder & Stoughton.

Nursall, J.R. 1981. Behaviour and habitat affecting the distribution of five species of sympatric mudskippers in Queensland. *Bull. mar. Sci.* **31**, 730–5.

O'Donnell, M.J. 1982. Water vapour absorption by the desert burrowing cockroach, *Arenivaga investigata*: evidence against a solute dependent mechanism. *J. exp. Biol.* **96**, 251–62.

Odum, E.P. 1971. *Fundamentals of ecology*. 3rd edn. Philadelphia: W.B. Saunders Co.

Oglesby, L.C. 1978. Salt and water balance. In *Physiology of annelids*, ed. P.J. Mill, pp. 555–658. London: Academic Press.

Okasha, A.Y.K. 1973. Water relations in an insect, *Thermobia domestica*. III. Effects of desiccation and rehydration on the haemolymph. *J. exp. Biol.* **58**, 385–400.

Oke, T.R. 1978. *Boundary layer climates*. London: Methuen.

Onuf, C.P., Teal, J.M. & Valiela, I. 1977. Interactions of nutrients, plant growth and herbivory in a mangrove ecosystem. *Ecology* **58**, 514–26.

Owen, J. 1980. *Feeding strategy*. Oxford University Press.

Packard, G.C. 1974. The evolution of air-breathing in paleozoic gnathostome fishes. *Evolution* **28**, 320–5.

Pampapathi Rao, K. 1963. Physiology of low temperature acclimation in tropical poikilotherms. I. Ionic changes in the blood of the freshwater mussel, *Lamellidens marginalis*, and the earthworm, *Lampito mauritii*. *Proc. Indian Acad. Sci.* **57B**, 290–6.

Panchen, A.L. 1980. The origin and relationships of the anthracosaur Amphibia from the late Palaeozoic. In *The terrestrial environment and the origin of land vertebrates*, ed. A.L. Panchen, pp. 319–50. London: Academic Press.

Panchen, A.L. & Smithson, T.R. 1987. Character diagnosis, fossils and the origin of tetrapods. *Biol. Rev.* **62**, 341–438.

Pantin, C.F.A. 1969. The genus *Geonemertes*. *Bull. Brit. Mus. nat. Hist.* **18**, 263–310.

Parry, D.A. 1951. Factors determining the temperature of terrestrial arthropods in sunlight. *J. exp. Biol.* **28**, 445–62.

Parry, D.A. 1954. On the drinking of soil capillary water by spiders. *J. exp. Biol.* **31**, 218–27.

Peaker, M. & Linzell, J.L. 1975. *Salt glands in birds and reptiles*. Cambridge University Press.

Pearse, A.S. 1936. *The migrations of animals from sea to land*. Durham, North Carolina: Duke University Press.

Pellegrino, C.R. 1984. The role of desiccation pressures and surface area/volume relationships on seasonal zonation and size distribution of four intertidal decapod Crustacea from New Zealand: implications for adaptation to land. *Crustaceana* **47**, 251–68.

Penteado, C.H.S. 1987. Respiratory responses of the tropical millipede *Plusioporus setiger* (Broelemann, 1902) (Spirostreptida: Spirostreptidae) to normoxic and hypoxic conditions. *Comp. Biochem. Physiol.* **86A**, 163–8.

Petersen, H. & Luxton, M. 1982. A comparative analysis of soil fauna populations and their role in decomposition processes. *Oikos* **39**, 287–388.

Pethick, J. 1984. *An introduction to coastal geomorphology*. London: Arnold.

Phillips, J.E. 1964. Rectal absorption in the desert locust, *Schistocerca gregaria* Forskal. I. Water. *J. exp. Biol.* **41**, 15–38.

Pierce, S.K. 1982. Invertebrate cell volume control mechanisms: a coordinated use of intracellular amino acids and inorganic ions as osmotic solute. *Biol. Bull. mar. biol. Lab., Woods Hole* **163**, 405–19.

Pilkington, J.B., Little, C. & Stirling, P.E. 1984. A respiratory current in the mantle cavity of *Amphibola crenata* (Mollusca, Pulmonata). *Jl R. Soc. N.Z.* **14**, 327–34.

Pollard, A.J. & Briggs, D. 1984. Genecological studies of *Urtica dioica* L. III. Stinging hairs and plant-herbivore interactions. *New Phytol.* **97**, 507–22.

Potts, W.T.W. 1954. The energetics of osmoregulation in brackish- and fresh-water animals. *J. exp. Biol.* **31**, 618–30.

1985. Discussion after 'The evolution of terrestrial vertebrates: environmental and physiological considerations' by A.A. Bray. *Phil Trans. R. Soc., Lond.* **B309**, 319–20.

Powers, L.W. & Bliss, D.E. 1983. Terrestrial adaptations. In *The biology of Crustacea*, eds. F.J. Vernberg & W.B. Vernberg, vol. 8, pp. 271–333. London: Academic Press.

Price, C.H. 1980. Water relations and physiological ecology of the salt marsh snail, *Melampus bidentatus* Say. *J. exp. mar. Biol. Ecol.* **45**, 51–67.

1984. Tidal migrations of the littoral salt marsh snail *Melampus bidentatus* Say. *J. exp. mar. Biol. Ecol.* **78**, 111–26.

Price, J.B. & Holdich, D.M. 1980. Changes in osmotic pressure and sodium concentration of the haemolymph of woodlice with progressive desiccation. *Comp. Biochem. Physiol.* **66A**, 297–305.

Prior, D.J., Hume, M., Varga, D. & Hess, S.D. 1983. Physiological and behavioural aspects of water balance and respiratory function in the terrestrial slug, *Limax maximus. J. exp. Biol.* **104**, 111–27.

Prosser, C.L. 1973. *Comparative animal physiology.* Philadelphia: Saunders.

Rackoff, J.S. 1980. The origin of the tetrapod limb and the ancestry of tetrapods. In *The terrestrial environment and the origin of land vertebrates,* ed. A.L. Panchen, pp. 255–92. London: Academic Press.

Rahn, H. & Howell, B.J. 1976. Bimodal gas exchange. In *Respiration of amphibious vertebrates,* ed. G.M. Hughes, pp. 271–85. London: Academic Press.

Ramsay, J.A. 1949. The osmotic relations of the earthworm. *J. exp. Biol.* **26**, 46–56.

1964. The rectal complex of the mealworm, *Tenebrio molitor* L. (Coleoptera, Tenebrionidae). *Phil. Trans. R. Soc. Lond.* **B248**, 279–314.

Randall, D.J., Burggren, W.W., Farrell, A.P. & Haswell, M.S. 1981. *The evolution of air breathing in vertebrates.* Cambridge University Press.

Rankin, J.C. & Davenport, J. 1981. *Animal osmoregulation.* Glasgow: Blackie.

Ranwell, D.S. 1972. *Ecology of salt marshes and sand dunes.* London: Chapman & Hall.

Raymond, A. & Phillips, T.L. 1983. Evidence for an Upper Carboniferous mangrove community. In *Biology and ecology of mangroves,* ed. H.J. Teas, pp. 19–30. The Hague: Dr W. Junk.

Reid, D.G. 1985. Habitat and zonation patterns of *Littoraria* species (Gastropoda: Littorinidae) in Indo-Pacific mangrove forests. *Biol. J. Linn. Soc,* **26**, 39–68.

1986. *The littorinid molluscs of mangrove forests in the Indo-Pacific region. The genus Littoraria.* London: British Museum (Natural History).

Reise, K. 1985. *Tidal flat ecology.* Berlin: Springer Verlag.

Remane, A. & Schlieper, C. 1971. *Biology of brackish water.* New York: John Wiley & Sons Inc.

Remmert, H. 1980. *Ecology.* Berlin: Springer-Verlag.

Retallack, G.J. 1985. Fossil soils as grounds for interpreting the advent of large plants and animals on land. *Phil. Trans. R. Soc.* **B309**, 105–42.

Retallack, G. & Dilcher, 1981. A coastal hypothesis for the dispersal and rise to dominance of flowering plants. In *Paleobotany, paleoecology and evolution,* vol. 2, New York: Praeger Publishers.

Reynoldson, T.B. 1966. The distribution and abundance of lake-dwelling triclads – towards a hypothesis. *Adv. ecol. Res.* **3**, 1–71.

1981. The ecology of the Turbellaria with special reference to the freshwater triclads. *Hydrobiologia* **84**, 87–90.

1983. The population biology of Turbellaria with special reference to the freshwater triclads of the British Isles. *Adv. Ecol. Res.* **13**, 235–26.

Richards, O.W. & Davies, R.G. 1977. *Imms' general textbook of entomology* (10th edition). London: Chapman & Hall.

Riddle, W.A. 1983. Physiological ecology of land snails and slugs. In *The Mollusca,* ed. W.D. Russell-Hunter, vol. 6, pp. 431–61. London: Academic Press.

Roach, D.K. 1963. Analysis of the haemolymph of *Arion ater* L. (Gastropoda: Pulmonata). *J. exp. Biol.* **40**, 613–23.

Robertson, A.I. 1986. Leaf-burying crabs: their influence on energy flow and export from mixed mangrove forests (*Rhizophora* spp.) in northeastern Australia. *J. exp. mar. Biol. Ecol.* **102**, 237–48.

Robison, R.A. 1985. Affinities of *Aysheaia* (Onychophora), with description of a new Cambrian species. *J. Paleontol.* **59**, 226–35.

Rolfe, W.D.I. 1980. Early invertebrate terrestrial faunas. In *The terrestrial environment and the origin of land vertebrates,* ed. A.L. Panchen, pp. 117–57. Systematics Association Special volume no. 15. London: Academic Press.

1985. Early terrestrial arthropods: a fragmentary record. *Phil. Trans. R. Soc.* **B309**, 207–18.

Romer, A.S. 1958. Tetrapod limbs and early tetrapod life. *Evolution* **12**, 365–9.

Root, R.B. 1973. Organization of a plant-arthropod association in simple and diverse habitats: the fauna of collards (*Brassica oleracea*). *Ecol. Monogr.* **43**, 95–124.

Rosen, D.E., Forey, P.L., Gardiner, B.G. & Patterson, C. 1981. Lungfishes, tetrapods, palaeontology, and plesiomorphy. *Bull. Am. Mus. nat. Hist.* **167**, 159–276.

Rumsey, T.J. 1972. Osmotic and ionic regulation in a terrestrial snail, *Pomatias elegans* (Gastropoda, Prosobranchia) with a note on some tropical Pomatiasidae. *J. exp. Biol.* **57**, 205–15.

Rumsey, T.J. 1973. Some aspects of osmotic and ionic regulation in *Littorina littorea* (L.) (Gastropoda, Prosobranchia). *Comp. Biochem. Physiol.* **45A**, 327–44.

Runham, N.W. & Hunter, P.J. 1970. *Terrestrial slugs*. London: Hutchinson.

Russell-Hunter, W.D., Apley, M.L. & Hunter, R.D. 1972. Early life-history of *Melampus* and the significance of semilunar synchrony. *Biol. Bull. mar. biol. Lab., Woods Hole* **143**, 623–56.

Salmon, M. & Hyatt, G.W. 1983. Communication. In *The biology of Crustacea*, eds. F.J. Vernberg & W.B. Vernberg, vol. 7, pp. 1–40. London: Academic Press.

Savory, T.H. 1971. *Evolution in the Arachnida*. Watford: Merrow.

Schaller, F. 1979. Significance of sperm transfer and formation of spermatophores in arthropod phylogeny. In *Arthropod phylogeny*, ed. A.P. Gupta, pp. 587–608. New York: Van Nostrand Rheinhold Co.

Schmalhausen, I.I. 1968. *The origin of terrestrial vertebrates*. London: Academic Press.

Schmidt-Nielsen, K. 1964. *Desert animals*. Oxford University Press.

1975. *Animal physiology*. Cambridge University Press.

Schmidt-Nielsen, K. Taylor, C.R. & Shkolnik, A. 1971. Desert snails: problems of heat, water and food. *J. exp. Biol.* **55**, 385–98.

Scotese, C.R., Van Der Voo, R. & Barrett, S.F. 1985. Silurian and Devonian base maps. *Phil. Trans. R. Soc.* **B309**, 57–77.

Seelemann, U. 1968. Zur Uberwindung der biologischen Grenze Meer-Land durch Mollusken. Untersuchungen an *Alderia modesta* (Opisth.) und *Ovatella myosotis* (Pulmonat.). *Oecologia* **1**, 130–54.

Selden, P. & Edwards, D. 1989. Colonization of the land. In *Evolution and ecology*, eds. K.C. Allen & D.E.G. Briggs. London: Pinter.

Selden, P.A. & Jeram, A.J. 1989. Palaeophysiology of terrestrialization in the Chelicerata. *Trans. R. Soc. Edinb., Earth Sciences,* **80**, 303–10.

Seymour, M.K. 1970. Skeletons of *Lumbricus terrestris* L. and *Arenicola marina* (L.). *Nature, London* **228**, 383–5.

Shaw, J. 1959. Salt and water balance in the east African fresh-water crab, *Potamon niloticus* (M.Edw.). *J. exp. Biol.* **36**, 157–76.

Shumway, S.E. & Freeman, R.F.H. 1984. Osmotic balance in a marine pulmonate, *Amphibola crenata. Mar. Behav. Physiol.* **11**, 157–83.

Singh, B.N. & Hughes, G.M. 1973. Cardiac and respiratory responses in the climbing perch, *Anabas testudineus. J. comp. Physiol.* **84**, 205–26.

Singh, B.N. & Munshi, J.S.D. 1968. On the respiratory organs and mechanics of breathing in *Periophthalmus vulgaris. Zool. Anz.* **183**, 92–110.

Slobodkin, L.B. & Rapoport, A. 1974. An optimal strategy of evolution. *Q. Rev. Biol.* **49**, 181–200.

Smith, H.W. 1959. *From fish to philosopher*. Boston: Little, Brown & Co.

Smith, W.K. & Miller, P.C. 1973. The thermal ecology of two south Florida crabs: *Uca rapax* Smith and *U. pugilator* Bosc. *Physiol. Zool.* **46**, 186–207.

Solem, A. 1985. Origin and diversification of pulmonate land snails. In *The Mollusca, vol. 10, Evolution*, eds. E.R. Trueman & M.R. Clarke, pp. 269–93. London: Academic Press.

Solem, A. & Yochelson, E.L. 1979. North American Paleozoic land snails, with a summary of other Paleozoic nonmarine snails. *Geol. Survey Prof. Pap.* **1072**, 1–42.

Southwood, T.R.E. 1973. The insect/plant relationship – an evolutionary perspective. *Symp. roy. ent. Soc. Lond.* **6**, 3–30.

1978. The components of diversity. In *Diversity of insect faunas*, eds. L.A. Mound & N. Waloff, pp. 19–40. *Symp. R. Ent. Soc. Lond.* **9**. Oxford: Blackwell.

1986. Plant surfaces and insects – an overview. In *Insects and the plant surface*, eds. B. Juniper & T.R.E. Southwood, pp. 1–22. London: Arnold.

Southwood, T.R.E. & Van Emden, H.F. 1967. A comparison of the fauna of cut and uncut grasslands. *Z. angew. Ent.* **60**, 188–98.

Spaargaren, D.H. 1978. A comparison of the blood osmotic composition of various marine and brackish-water animals. *Comp. Biochem. Physiol.* **60**. 327–33.

Speeg, K.V. & Campbell, J.W. 1968. Formation and volatilization of ammonia gas by terrestrial snails. *Am. J. Physiol.* **214**, 1392–402.

Spencer, J.O. & Edney, E.B. 1954. The absorption of water by woodlice. *J. exp. Biol.* **31**, 491–6.

Spicer, J.I., Moore, P.G. & Taylor, A.C. 1987. The physiological ecology of land invasion by the Talitridae (Crustacea: Amphipoda). *Proc. R. Soc.* **B232**, 95–124.

Spicer, J.I. & Taylor, A.C. 1987. Respiration in air and water of some semi- and fully terrestrial talitrids (Crustacea: Amphipoda: Talitridae). *J. exp. mar. Biol. Ecol.* **106**, 265–77.

Spicer, R.A. 1989. Physiological characteristics of land plants in relation to environment through time. *Trans. roy. Soc. Edinb., Earth Sciences,* **80**, 321–9.

Stahl, B.J. 1974. *Vertebrate history: problems in evolution.* New York: McGraw-Hill.

Stebbins, G.L. & Hill, G.J.C. 1980. Did multicellular plants invade the land? *Am. Nat.* **165**, 342–53.

Stebbins, R.C. & Kalk, M. 1961. Observations on the natural history of the mud-skipper, *Periophthalmus sobrinus. Copeia* (1961) 18–27.

Steinberger, Y., Grossman, S. & Dubinsky, Z. 1981. Some aspects of the ecology of the desert snail *Sphincterochila prophetarum* in relation to energy and water flow. *Oecologia* **50**, 103–8.

Steinberger, Y., Grossman, S., Dubinsky, Z. & Shachak, M. 1983. Stone microhabitats and the movement and activity of desert snails, *Sphincterochila prophetarum. Malacol. Rev.* **16**, 63–70.

Stephenson, J. 1930. *The Oligochaeta.* Oxford: Clarendon Press.

Stephenson, T.A. & Stephenson, A. 1972. *Life between tidemarks on rocky shores.* San Francisco: W.H. Freeman & Co.

Stewart, W.N. 1983. *Paleobotany and the evolution of plants.* Cambridge University Press.

Stobbart, R.H. & Shaw, J. 1974. Salt and water balance: excretion. In *The physiology of Insecta* (2nd edition), ed. M. Rockstein, vol. 5, pp. 362–446. London: Academic Press.

Strong, D.R. 1984. Exorcising the ghost of competition past: phytophagous insects. In *Ecological communities. Conceptual issues and the evidence*, eds. D.R. Strong, D. Simberloff, L.G. Abele & A.B. Thistle, pp. 28–41. Princeton, New Jersey: Princeton University Press.

Strong, D.R., Lawton, J.H. & Southwood, T.R.E. 1984. *Insects on Plants. Community patterns and mechanisms.* Oxford: Blackwell.

Sundnes, G. & Valen, E. 1969. Respiration of dry cysts of *Artemia salina* L. *J. Cons. perm. int. Explor. Mer* **32**, 413–15.

Sutton, S.L. & Holdich, D.M. (eds.) 1984. The biology of terrestrial isopods. *Symp. Zool. Soc. Lond.* **53**, 1–518.

Sverdrup, H.U., Johnson, M.W. & Fleming, R.H. 1942. *The oceans.* Englewood Cliffs, New Jersey: Prentice-Hall, Inc.

Sykes, A.H. 1971. Formation and composition of urine. In *Physiology and biochemistry of the domestic fowl*, eds. D.J. Bell & B.M. Freeman: vol. I, pp. 233–78. London: Academic Press.

Szarski, H. 1976. Sarcopterygii and the origin of tetrapods. In *Major patterns in vertebrate evolution*, ed. M.K. Hecht, P.C. Goody & B.M. Hecht, pp. 517–40. London: Plenum Press.

Takeuchi, N. 1980a. A possibility of elevation of the free amino acid level for the extra-cellular hyperosmotic adaptation of the earthworm *Eisenia foetida* (Sav.) to the concentrated medium. *Comp. Biochem. Physiol.* **67A**, 353–5.

1980b. Control of coelomic fluid concentration and brain neurosecretion in the littoral earthworm *Pontodrilus matsushimensis* Iizuka. *Comp. Biochem. Physiol.* **67A**, 357–9.

Tamura, H. & Koseki, K. 1974. Population study on a terrestrial amphipod, *Orchestia platensis japonica* (Tattelsall), in a temperate forest. *Jap. J. Ecol.* **24**, 123–39.

Taylor, A.C. & Spicer, J.I. 1986. Oxygen-transporting properties of the blood of two semi-terrestrial amphipods, *Orchestia gammarellus* (Pallas) and *O. mediterranea* (Costa). *J. exp. mar. Biol. Ecol.* **97**, 135–50.

Taylor, E.W. & Innes, A.J. 1988. A functional analysis of the shift from gill- to lung-breathing during the evolution of land crabs (Crustacea, Decapoda). *Biol. J. Linn. Soc.* **34**, 229–47.

Teal, J.M. 1959. Respiration of crabs in Georgia salt marshes and its relation to their ecology. *Physiol. Zool.* **32**, 1–14.

1962. Energy flow in the salt marsh ecosystem of Georgia. *Ecology* **43**, 614–24.

Teal, J.M. & Teal, M. 1969. *Life and death of the salt marsh.* New York: Audubon/Ballantine.

Thompson, F.G. 1980. Proserpinoid land snails and their relationships within the Archaeogastropoda. *Malacologia* **20**, 1–33.

Thomson, K.S. 1980. The ecology of the Devonian lobe-finned fish. In *The terrestrial environment and the origin of land vertebrates*, ed. A.L. Panchen, pp. 187–222. Systematics Association Special Volume no. 15. London: Academic Press.

Timm, T. 1980. Distribution of aquatic oligochaetes. In *Aquatic oligochaete biology*, eds. R.O. Brinkhurst & D.G. Cook, pp. 55–7. London: Plenum Press.

Townsend, C.R. 1980. *The ecology of streams and rivers.* London: Arnold.

Trueman, E.R. 1975. *The locomotion of soft-bodied animals.* London: Arnold.

Underwood, A.J. 1978. A refutation of critical tidal levels as determinants of the structure of intertidal communities on British shores. *J. exp. mar. Biol. Ecol.* **33**, 261–76.

1979. The ecology of intertidal gastropods. *Adv. mar. Biol.* **16**, 111–210.

Underwood, A.J., Denley, E.J. & Moran, M.J. 1983. Experimental analyses of the structure and dynamics of mid-shore rocky intertidal communities in New South Wales. *Oecologia (Berlin)* **56**, 202–19.

Underwood, A.J. & McFadyen, K.E. 1983. Ecology of the intertidal snail *Littorina acutispira* Smith. *J. exp. mar. Biol. Ecol.* **66**, 169–97.

Usher, M.B. 1985. Population and community dynamics in the soil ecosystem. In *Ecological interactions in soil*, eds. A.H. Fitter, D. Atkinson, D.J. Read & M.B. Usher, pp. 243–65. Oxford: Blackwell.

Usher, M.B., Davis, P.R., Harris, J.R.W. & Longstaff, B.C. 1979. A profusion of species? Approaches towards understanding the dynamics of the populations of the micro-arthropods in decomposer communities. In *Population dynamics*, eds. R.M. Anderson, B.D. Turner & L.R. Taylor, pp. 359–84. Oxford: Blackwell.

Valentine, J.W. & Moores, E.M. 1976. Plate tectonics and the history of life in the oceans. In *Continents adrift and continents aground*, ed. J.T. Wilson: pp. 196–205. San Francisco: W.H. Freeman.

Vandel, A. 1965. Sur l'existence d'oniscoides tres primitifs menant une vie aquatique et sur le polyphyletisme des isopodes terrestres. *Ann. Speleologie* **20**, 489–518.

Vannier, G. 1983. The importance of ecophysiology for both biotic and abiotic studies of the soil. In *New trends in soil biology*, eds. P. Lebrun, H.M. Andre, A. de Medts, C. Gregoire-Wibo & G. Wauthy, pp. 289–314. Proc. VIII Int. Colloquium of Soil Zoology. Ottignies-Louvain-la-Neuve: Dieu-Brichart.

Varley, G.C., Gradwell, G.R. & Hassell, M.P. 1973. *Insect population ecology. An analytical approach*. Oxford: Blackwell.

Vermeij, G.J. 1971. Temperature relationships of some tropical Pacific intertidal gastropods. *Mar. Biol.* **10**, 308–14.

Vernberg, F.J. 1984. Fiddler crabs: ecosystems – organisms – molecules. *Am. Zool.* **24**, 293–304.

Vorwohl, G. 1961. Zur Funktion der Exkretionsorgane von *Helix pomatia* L. und *Archachatina ventricosa* Gould. *Z. vergl. Physiol.* **45**, 12–49.

Wall, B.J. 1977. Fluid transport in the cockroach rectum. In *Transport of ions and water in animals*, eds. B.L. Gupta, R.B. Moreton, J.L. Oschman & B.J. Wall, pp. 599–612. London: Academic Press.

Wallwork, J.A., 1976. *The distribution and diversity of soil fauna*. London: Academic Press.

Walsh, G.E. 1974. Mangroves: a review. In *Ecology of halophytes*, eds. R.J. Reimold & W.H. Queen, pp. 51–174. London: Academic Press.

Warburg, M.R. 1987. Isopods and their terrestrial environment. *Adv. ecol. Res.* **17**, 187–242.

Warner, G.F. 1967. The life history of the mangrove tree crab, *Aratus pisoni*. *J. Zool. (Lond.)* **153**, 321–35.

1969. The occurrence and distribution of crabs in a Jamaican mangrove swamp. *J. anim. Ecol.* **38**, 379–89.

Warren, J.H. 1985. Climbing as an avoidance behaviour in the salt marsh periwinkle, *Littorina irrorata* (Say). *J. exp. mar. Biol. Ecol.* **89**, 11–28.

Warren, J.H. & Underwood, A.J. 1986. Effects of crabs on the topography of mangrove swamps in New South Wales. *J. exp. mar. Biol. Ecol.* **102**, 223–35.

Waterhouse, F.L. 1955. Microclimatological profiles in grass cover in relation to biological problems. *Q. J. roy. met. Soc.* **81**, 63–71.

Watson, D.M.S. 1951. *Paleontology and modern biology*. New Haven: Yale University Press.

Webb, J.E. 1957. The ecology of Lagos lagoon V. Some physical properties of lagoon deposits. *Phil. Trans. R. Soc.* B, **241**, 393–419.

Weigman, G. 1973. Zur Okologie der Collembolen und Oribatiden im Grenzbereich Land-Meer (Collembola, Insecta – Oribatei, Acari). *Z. wiss. Zool.* **186**, 295–391.

Wells, F.E. 1984. Comparative distribution of macromolluscs and macrocrustaceans in a north-western Australian mangrove system. *Aust. J. mar. fw. Res.* **35**, 591–6.

Whittington, H.B. 1978. The lobopod animal *Aysheaia pedunculata* Walcott, Middle Cambrian, Burgess Shale, British Columbia. *Phil. Trans. R. Soc.* **B284**, 165–97.

1985. *The Burgess shale*. New Haven: Yale University Press.

Wieser, W. & Schuster, M. 1975. The relationship between water content, activity, and free amino acids in *Helix pomatia* L. *J. Comp. Physiol.* **98**, 169–81.

Wieser, W. & Schweizer, G. 1970. A re-examination of the excretion of nitrogen by terrestrial isopods. *J. exp. Biol.* **52**, 267–74.

Wieser, W., Schweizer, G. & Hartenstein, R. 1969. Patterns of release of gaseous ammonia by terrestrial isopods. *Oecologia* **3**, 390–400.

Wildish, D.J. 1970. Locomotory activity rhythms in some littoral *Orchestia* (Crustacea: Amphipoda). *J. mar. Biol. Ass. UK* **50**, 241–52.

Williams, B.G., Naylor, E. & Chatterton, T.D. 1985. The activity patterns of New Zealand mud crabs under field and laboratory conditions. *J. exp. mar. Biol. Ecol.* **89**, 269–82.

Williams, E.C. 1941. An ecological study of the floor fauna of the Panama rain forest. *Bull. Chicago Acad. Sci.* **6**, 63–124.

Williams, G.A. 1989. *Littorina mariae* – a factor structuring low shore communities? *Hydrobiologia*, in press.

Williams, J.A. 1979. A semi-lunar rhythm of locomotory activity and moult synchrony in the sand-beach amphipod *Talitrus saltator*. In *Cyclic phenomena in marine plants and animals*, eds. E. Naylor & R. Hartnoll, pp. 407–14. Oxford: Pergamon Press.

    1982. A circadian rhythm of oxygen consumption in the sand beach amphipod *Talitrus saltator* (Montagu). *J. exp. mar. Biol. Ecol.* **57**, 125–34.

Williamson, D.I. 1951. On the mating and breeding of some semi-terrestrial amphipods. *Rep. Dove mar. Lab* (Ser. 3) **12**, 49–62.

Willmer, P. 1982. Microclimate and the environmental physiology of insects. *Adv. Insect Physiol.* **16**, 1–57.

    1986. Microclimatic effects on insects at the plant surface. In *Insects and the plant surface*, eds. B. Juniper & T.R.E. Southwood, pp. 65–80. London: Arnold.

Wilson, K. 1960. The time factor in the development of dune soils at South Haven Peninsula, Dorset. *J. Ecol.* **48**, 341–59.

Wilson, R.A. & Webster, L.A. 1974. Protonephridia. *Biol. Rev.* **49**, 127–60.

Wilson, W.J. 1970. Osmoregulatory capabilities in isopods: *Ligia occidentalis* and *Ligia pallasii*. *Biol. Bull. mar. biol. Lab., Woods Hole* **138**, 96–108.

Wise, D.H. 1984. The role of competition in spider communities: insights from field experiments with a model organism. In *Ecological communities. Conceptual issues and the evidence*, eds. D.R. Strong, D. Simberloff, L.G. Abele & A.B. Thistle, pp. 42–53. Princeton, New Jersey: Princeton University Press.

Wolcott, T.G. 1976. Uptake of soil capillary water by ghost crabs. *Nature, Lond.* **264**, 756–7.

    1984. Uptake of interstitial water from soil: mechanisms and ecological significance in the ghost crab *Ocypode quadrata* and two gecarcinid land crabs. *Physiol. Zool.* **57**, 161–84.

Wolcott, T.G. & Wolcott, D.L. 1985. Factors influencing the limits of migratory movements in terrestrial crustaceans. *Contr. mar. Sci.* **27**, 257–73.

    1988. Availability of salts is not a limiting factor for the land crab *Gecarcinus lateralis* (Freminville). *J. exp. mar. Biol. Ecol.* **120**, 199–219.

Wood, C.M. & Caldwell, F.H. 1978. Renal regulation of acid-base balance in a freshwater fish. *J. exp. Zool.* **205**, 301–7.

Wood, T.G. 1971. The distribution and abundance of *Folsomides deserticola* (Collembola: Isotomidae) and other microarthropods in arid and semi-arid soils in southern Australia, with a note on nematode populations. *Pedobiologia* **11**, 446–68.

Wootton, R.J. 1979. Energy costs of egg production and environmental determinants of fecundity in teleost fishes. *Symp. zool. Soc. Lond.* **44**, 133–59.

Wright, D.A., Zanders, I.P. & Pait, A. 1984. Ionic regulation in three species of *Uca*: a comparative study. *Comp. Biochem. Physiol.* **78A**, 175–9.

Wright, V.P. 1985. The precursor environment for vascular plant colonization. *Phil. Trans. R. Soc.* **B309**, 143–5.

Young, J.O. 1981. A comparative study of the food niches of lake-dwelling triclads and leeches. *Hydrobiologia* **84**, 91–102.

# Index